FROM MONSOONS TO MICROBES

Understanding the Ocean's Role in Human Health

Committee on the Ocean's Role in Human Health

Ocean Studies Board

Commission on Geosciences, Environment, and Resources

National Research Council

NATIONAL ACADEMY PRESS
Washington, D.C.   1999

NATIONAL ACADEMY PRESS • 2101 Constitution Ave., N.W. • Washington, D.C. **20418**

NOTICE: The project that is the subject of this report was approved by the Governing Board of the National Research Council, whose members are drawn from the councils of the National Academy of Sciences, the National Academy of Engineering, and the Institute of Medicine. The members of the committee responsible for the report were chosen for their special competencies and with regard for appropriate balance.

This report and the committee were supported by grants from the National Oceanic and Atmospheric Administration, the National Institute of Environmental Health Sciences, and the National Aeronautics and Space Administration. The views expressed herein are those of the authors and do not necessarily reflect the views of the sponsors.

Library of Congress Cataloging-in-Publication Data

From monsoons to microbes : understanding the ocean's role in human
health / Ocean Studies Board, Commission on Geosciences,
Environment, and Resources, National Research Council.
    p. cm.
    Includes bibliographical references (p.   ) and index.
    ISBN 0-309-06569-0 (casebound)
    1. Marine pollution—Health aspects. 2. Marine microbiology. 3.
Marine pharmacology. I. National Research Council (U.S.). Ocean
Studies Board.
    RA600 .F76 1999
    616.9'8—dc21   99-6094

From Monsoons to Microbes: Understanding the Ocean's Role in Human Health is available from the National Academy Press, 2101 Constitution Ave., N.W., Lockbox 285, Washington, DC 20055; (800) 624-6242 OR (202) 334-3313 (in the Washington metropolitan area); Internet, http://www.nap.edu

Cover art: Small single-celled algae known as dinoflagellates fall on the background of a rainstorm, along with a satellite image of a hurricane brewing over the tropical ocean. The dinoflagellates are watercolors taken from C.A. Kofoid and O. Swezy (1921), "The Free-Living Unarmored Dinoflagellata," *Memoirs of the University of California*, Vol. 5, University of California Press, Berkeley.

COMMITTEE ON THE OCEAN'S ROLE IN HUMAN HEALTH

WILLIAM FENICAL, *Chair*, Scripps Institution of Oceanography, La Jolla, California
DANIEL BADEN, University of Miami, Miami, Florida
MAURICE BURG, National Institutes of Health, Bethesda, Maryland
CLAUDE DE VILLE DE GOYET, Pan American Health Organization, Washington, D.C.
DARRELL JAY GRIMES, The University of Southern Mississippi, Ocean Springs
MICHAEL KATZ, March of Dimes, White Plains, New York
NANCY MARCUS, Florida State University, Tallahassee
SHIRLEY POMPONI, Harbor Branch Oceanographic Institution, Inc., Fort Pierce, Florida
PETER RHINES, University of Washington, Seattle
PATRICIA TESTER, National Marine Fisheries Service, NOAA, Beaufort, North Carolina
JOHN VENA, University at Buffalo, State University of New York, Buffalo

Staff
SUSAN ROBERTS, Study Director
SHARI MAGUIRE, Senior Project Assistant

The National Academy of Sciences is a private, nonprofit, self-perpetuating society of distinguished scholars engaged in scientific and engineering research, dedicated to the furtherance of science and technology and to their use for the general welfare. Upon the authority of the charter granted to it by the Congress in 1863, the Academy has a mandate that requires it to advise the federal government on scientific and technical matters. Dr. Bruce M. Alberts is president of the National Academy of Sciences.

The National Academy of Engineering was established in 1964, under the charter of the National Academy of Sciences, as a parallel organization of outstanding engineers. It is autonomous in its administration and in the selection of its members, sharing with the National Academy of Sciences the responsibility of advising the federal government. The National Academy of Engineering also sponsors engineering programs aimed at meeting national needs, encourages education and research, and recognizes the superior achievements of engineers. Dr. William A. Wulf is president of the National Academy of Engineering.

The Institute of Medicine was established in 1970 by the National Academy of Sciences to secure the services of eminent members of appropriate professions in the examination of policy matters pertaining to the health of the public. The Institute acts under the responsibility given to the National Academy of Sciences by its congressional charter to be an adviser to the federal government and, upon its own initiative, to identify issues of medical care, research, and education. Dr. Kenneth I. Shine is president of the Institute of Medicine.

The National Research Council was organized by the National Academy of Sciences in 1916 to associate the broad community of science and technology with the Academy's purposes of furthering knowledge and advising the federal government. Functioning in accordance with general policies determined by the Academy, the Council has become the principal operating agency of both the National Academy of Sciences and the National Academy of Engineering in providing services to the government, the public, and the scientific and engineering communities. The Council is administered jointly by both Academies and the Institute of Medicine. Dr. Bruce M. Alberts and Dr. William A. Wulf are chairman and vice-chairman, respectively, of the National Research Council.

Foreword

1998 has been declared the International Year of the Ocean (YOTO). This has led to the initiation of a number of activities meant to enhance the public's awareness of the ocean and to improve our ability to deal effectively with the hydrosphere. One useful and important activity that each of us could undertake as YOTO draws to a close would be to think about the ways the ocean affects our lives. Some connections are clear: people in the transportation industry might express concern about tides, winds and currents and how they affect the safety and economics of shipping. People in the fishing industry might recognize how their livelihoods depend on the health and productivity of the fishing grounds. The broader public might value the ocean as a source of food and recreation, and remember how the warm Pacific Ocean waters of the 1997-98 El Niño brought unusually warm, wet weather to much of the United States. There are many other immediate connections between the ocean and human activities.

However, this report examines another, less often recognized, aspect of how the ocean affects our lives; the implications of ocean phenomena for human health. That this issue has not been discussed broadly is probably a reflection of the diverse ways in which the ocean influences health. The following report explores the nature of these connections, considers the state of knowledge in important areas, and makes recommendations for how improvements can be made in human health through a better understanding of the oceans.

KENNETH BRINK
Chair, Ocean Studies Board

Preface

The Committee on the Ocean's Role in Human Health was charged with examining a variety of ways in which the oceans play a role in human health: from large-scale physical processes to micro-scale biochemical processes. This report is intended as an overview of these issues, a starting point for considering how the marine sciences have contributed and can continue to contribute to improving human health.

This study began with a workshop on the Ocean's Role in Human Health to bring together members of the ocean sciences, medical, and public health communities for discussion of various topics connecting the study of marine processes and marine organisms to the promotion of human health. The committee extends its gratitude to the following individuals who spoke at the workshop and provided background information for the report: Lorraine Backer, Robert Baker, Frances Carr, David Epel, Joan Ferraris, Sherwood Hall, Anwarul Huq, John Marchalonis, Baldomero Olivera, Joan Rose, Lynn "Nick" Shay, Benjamin Sherman, Erika Siegfried, Milan Trpis, and William Wiseman.

The committee is grateful for the assistance provided by the following individuals who provided additional background material, data, and figures for consideration and use by the committee: Donald M. Anderson, Paul Epstein, Eric L. Geist, George N. Pavlakis, Lynn "Nick" Shay, and Stephen A. Stricker. For their assistance in data gathering, preparation, and consultation the committee extends its thanks to the following individuals: Constance Carter and Adrienne Davis.

This report has been reviewed in draft form by individuals chosen for their diverse perspectives and technical expertise, in accordance with procedures approved by the NRC's Report Review Committee. The purpose of this independent review is to provide candid and critical comments that will assist the institution in making the published report as sound as possible and to ensure that the report meets institutional standards for objectivity, evidence, and responsiveness to the study charge. The review comments and draft manuscript remain confidential to protect the integrity of the deliberative process. We wish to thank the following individuals for their participation in the review of this report: Duane Gubler, Centers for Disease Control and Prevention; Judith McDowell, Woods Hole Oceanographic Institution; Jonathan Patz, Johns Hopkins School of Public Health; Roger Pielke, National Center for Atmospheric Research; Michael Roman, University of Maryland; Sandra Shumway, Southampton College, Long Island University; Patrick Walsh, University of Miami; and Jaw-Kai Wang, University of Hawaii. While the individuals listed above have provided constructive comments and suggestions, it must be emphasized that responsibility for the final content of this report rests entirely with the authoring committee and the institution.

The committee gratefully acknowledges the efforts of the Ocean Studies Board (OSB) staff who helped to produce this report, particularly the study director, Susan Roberts, and the project assistant, Shari Maguire. For their efforts in bringing this activity to fruition, the committee wishes to thank Morgan Gopnik, OSB director; Daniel Walker, OSB program officer; and the staff of the Board on Health Sciences Policy at the Institute of Medicine. This study was funded by the National Oceanic and Atmospheric Administration, the National Institute of Environmental Health Sciences, and the National Aeronautics and Space Administration.

WILLIAM FENICAL
Chair, Committee on the Ocean's Role in Human Health

Contents

xi

Executive Summary

The United Nations declared 1998 the International Year of the Ocean. Activities during the past year have provided a renewed appreciation for the resources obtained from the ocean, the effects of human activities on the health of the ocean, and the importance of the ocean in regulating the world's climate. The more we learn about ocean processes and ocean life, the more we realize how critical the ocean will be for the future well-being of humankind. Human health is one of the areas strongly influenced by the ocean. There are negative impacts such as the spread of infectious diseases, coastal weather hazards, and harmful algal blooms, and positive impacts such as the use of marine organisms to develop new medical treatments and a better understanding of biological processes.

In recognition of the International Year of the Ocean, the National Research Council held a workshop on the Ocean's Role in Human Health in June, 1998. The workshop brought together members of the ocean sciences and biomedical communities to identify areas where improved understanding of marine processes and systems has the potential to reduce public health risks and enhance our existing biomedical capabilities. This document serves as an overview, based partly on discussions at the workshop, of the impacts of the ocean on human health with emphasis on (1) elucidating connections between the ocean and human health, (2) evaluating the present state of knowledge about these connections, and (3) suggesting how current and future efforts may be directed so that we can anticipate and respond to future health needs and threats.[1]

---

[1] Because of the complex nature of both human health and ocean processes, and the limited scope of this study, the related issues of secondary effects of pollution and the contribution of the ocean to the world's food supply are not examined in this report.

CONNECTIONS BETWEEN THE OCEAN AND HUMAN HEALTH

Marine Processes That Threaten Public Health

The ocean acts as a conduit for many human diseases. The distribution of viral, bacterial, and protozoal agents and algal toxins in marine habitats depends on the interplay of currents, tides, and human activities. The primary route of human exposure is through ingestion of contaminated seafood, but illness can also result from direct contact with seawater during recreational or occupational activities and from contact through aerosols (sea spray) containing toxins.

Disease-causing organisms can be spread by several different marine processes. Coastal and estuarine circulation patterns influence the frequency and geographic pattern of harmful algal blooms. Nutrient loading from heavy run-off also poses problems of anoxia and contributes to the proliferation of algae. Also, circulation of waters through estuaries and coastal areas plays a role in determining where and when the risks of contamination by human pathogens are highest. Pathogens from human or animal waste contaminate coastal and estuarine areas through freshwater runoff from sewers, rivers, and streams. Viruses (e.g., hepatitis A and poliovirus) and bacteria (*E. coli* and *Salmonella*) of fecal origin become concentrated in filter-feeding shellfish such as oysters and clams. Marine pathogenic bacteria (e.g., *Vibrio cholera*) and harmful algal species can invade new areas through the transport of organisms in the ballast water of ships. Thus shipping activities can introduce a disease from one part of the world into a new location causing ecological, economic, and human health problems. Harmful algal species can also be transported great distances by major ocean currents such as the Gulf Stream. Finally, international trade transmits algal toxins and pathogens through commerce in seafood.

In addition to these specific health threats from infectious and toxic organisms, there are both seasonal and periodic public health concerns that arise from severe weather and climate variability. Climate and weather are determined through interdependent atmospheric and oceanic processes. The world ocean functions as a huge reservoir of heat and moisture that fuels weather systems. These weather systems in turn affect the ocean by wind-driven mixing of surface and deep waters and by changes in sea level dependent on barometric pressure, winds, and melting of polar ice caps during warm periods.

The most vivid and direct impacts of the ocean on human health arise in coastal areas that are subject to tsunamis, storm surges, heavy rainfall and flooding, and severe winds. High water associated with torrential rains, storm surges, and tsunamis result in the highest mortality. However, lingering economic losses from damage caused by severe weather like a tropical storm can cause an overall decrease in public health in developing countries through increased poverty and the loss of housing, hospitals, and public sanitation systems.

The incidence and intensity of severe weather systems are affected by the recurring climatic patterns known as the El Niño / Southern Oscillation (ENSO) and the North Atlantic Oscillation (NAO). These phenomena originate from shifts in the temperature and flow of ocean water masses and affect global weather systems, bringing droughts to some regions and torrential storms to others. In addition to the direct effects on coastal areas described above, these weather patterns can have impacts on inland areas through shifts in temperature and rainfall patterns which affect the density and distribution of disease-causing organisms.

The observed increase in global average temperature over the past century has generated concern that the world may be entering a period of global warming in excess of normal climatic variation (Nicholls et al., 1995). Such a change in the heat balance of the ocean-atmosphere system could lead to dramatic effects on climate and local weather patterns. This has the potential to affect the frequency and severity of tropical storms and periodic events like ENSO (Gray, 1984; Gray et al., 1993; Gray et al., 1994; Landsea et al., 1994). Also, shifts in temperature and rainfall patterns influence the distribution of organisms that cause human diseases, including harmful algae, waterborne agents such as the pathogenic vibrios, and vectors of disease such as mosquitoes that carry malaria, dengue fever, and yellow fever. There may be public health effects due to unusual temperature extremes, air pollution, and the availability of fresh water and food.

Contributions of Marine Biodiversity to Biomedicine

In addition to recognizing the health problems associated with the ocean, this report describes how the ocean provides society with an essential biomedical resource through the rich diversity of marine organisms. Plants, animals, and microbes have provided either the source or the concept for more than half of the pharmaceuticals currently on the market. The emphasis has traditionally been on using terrestrial organisms, but despite continued and more sophisticated searches for new bioactive agents, there has been a decreasing return in molecular diversity, and hence new drug compounds. At the same time, many of the bacteria that cause life-threatening diseases have become resistant to existing antibiotics, making the need for new drug discovery more urgent. Also, new bioactive agents are needed to overcome the limitations of our current arsenal of drugs for treating cancer and other diseases. Because the diversity of life at higher taxonomic levels is greater in the ocean than on land, marine organisms offer a promising source of novel compounds with therapeutic potential. A number of these compounds are already under investigation or development by the pharmaceutical industry for the treatment of diseases such as cancer. Nevertheless, there are significant challenges to be surmounted in collecting organisms from marine environments and uncovering new compounds with potential value as pharmaceuticals. Meeting these challenges will require cooperation and new programs that bring together scientists from both the oceanographic and biomedical communities.

Marine biodiversity also benefits medical research because scientists use marine organisms as models for basic research on biological and disease processes. Marine organisms are valued as experimental models for studies of essential molecular, cellular, and physiological processes. The diversity of oceanic life helps scientists investigate and understand the evolutionary basis of many fundamental processes such as the biochemical basis of learning and memory, the mechanics of cell division, and the physiology of salt tolerance in the kidney. Also, unique properties found in marine organisms facilitate the investigation of complex biological processes. For example, scientists have exploited the unusually large diameter of the giant axon in a squid neuron to study the electrophysiology of nerve impulses. In addition, marine animal models have provided insights into the origin of human diseases such as diabetes and cancer.

CONCLUSIONS

In this section we present three approaches to further our understanding of the ocean's role in human health that cross-cut the specific issues described in each chapter.

I. Information Resources for Improved Prediction and Prevention of Marine Public Health Disasters

Prediction and prevention of public health disasters precipitated by ocean phenomena depend on establishing an historical baseline of high quality observations. Predictions of marine events that affect health depend critically on the quality of the data used to develop and test models. To examine recurring and long term climate variations, observations need to be collected regularly over periods long enough to distinguish patterns. This requires an ongoing commitment to monitoring and analysis from both funding agencies and scientists. The advent of automated data gathering of physical and biological measurements has created the requirement for comprehensive, structured databases. Also, access to the databases must include query-driven retrieval systems. Through the establishment of baselines and documentation of trends, these systems will help researchers evaluate the implications of an ecological disturbance or a climatological event.

Priorities for monitoring programs

The following activities are important for gathering information needed to address the health issues discussed in this report:

1. Collection of baseline observations of physical ocean properties to monitor climate variation on a global scale.

Climate change represents not only a variation in global average temperature, but also a patchwork pattern of change in temperature, severity of storms, rainfall and drought, ocean circulation, and upwelling frequency. All of these changes will affect the ecosystems on which many human communities depend.

2. Measurement of oceanic (e.g., upper ocean heat content) and atmospheric (e.g., jet stream level winds, storm core dynamics) variables to improve tropical storm predictions.

Although disastrous events such as tropical storms cannot be prevented, improved predictions can help reduce the costs. It is estimated that the expense of evacuating a coastal area in the southern U.S. in preparation for a hurricane is approaching $1M per mile of coast (OFCM, 1997). Hence better predictions of the site of landfall could dramatically reduce the expense of the storm and the disruption of coastal communities.

3. Collection of an organized, ongoing compilation of health statistics, particularly in developing countries that lack extensive public health infrastructure, to allow retrospective analysis of the effects of oceanic events on human health.

Reduction and prevention of human health threats from oceanic phenomena requires determination of cause and effect, which is possible only by correlating oceanographic and atmospheric data with reliable reporting from the public health sector. First, it is important that infectious diseases and algal toxin poisonings be correctly diagnosed. Second, information on the frequency, location, and date of disease outbreaks needs to be accurately reported and made available to the international health community. Finally, the efficacy and comprehensiveness of disease surveillance programs need to be evaluated. Associations between environmental events such as El Niño and changes in human disease patterns will require an historical perspective. Participation by experts in many fields and by the appropriate international agencies will be needed to document, evaluate, and develop mitigation strategies. Professional societies in the disciplines of microbiology, meteorology, oceanography, parasitology, and epidemiology should be involved in the establishment of research priorities and methodological development.

4. Documentation of harmful algal blooms (HABs)

The reports of HABs have been increasing, with higher frequency of occurrence and greater geographic range of several species since the early 1970s. Accurate species identification is needed during a bloom to track species dispersal and identify toxin hazards. Increased monitoring of water conditions will allow comparisons before, during, and after an algal bloom to determine the physical, chemical, and biological factors that promote blooms of specific harmful algal species. Effective mitigation efforts will depend on the identification of the causes of the increasing incidence and geographic distribution of HABs.

II. New Technological Approaches to Help Reduce Risks to Human Health

New technologies offer greater opportunities to explore and understand the ocean and marine life. Satellite oceanography and molecular probes (to monitor both ocean chemistry and the contamination of coastal waters by pathogens or algal toxins) provide more detailed information about marine environments than previous methods. Piloted and remotely operated submersible vehicles make it possible to study and retrieve new species from the underexplored ocean floor. Although great advances have been made in their development, implementation of these new technologies has not yet been fully realized.

Priorities for implementation

Several of the technologies that could help address current and future health issues include:

1. Drifting and moored sensors (monitored by satellite), although not a new technology, need to be deployed more widely to accurately track ocean processes. Similarly, continued submersible development is needed to study and collect deep sea organisms.

2. Biological sensors will improve *in situ* measurements of biological processes in marine coastal waters and allow higher sensitivity measurements of water quality (including specific nutrients, oxygen, pH, species-specific monitoring of algae and bacteria).

3. DNA probes and antibody-based tests will provide more sensitive and specific detection of pathogens in marine waters as a supplement to the current standard, the coliform test, which acts as an indirect indicator of fecal pollution (i.e., sewage).

4. More accurate, cost-effective methods for detecting algal toxins in seafood, based on the molecular properties of the toxins.

III. Contributions of Marine Organisms to Medicine and Research

The diversity of life in the ocean has the potential to contribute to the development of effective new treatments of human diseases and a greater understanding of human biology. However, the level of effort expended to use this marine resource for biomedical applications has been modest compared to the potential benefits (NRC, 1994a).

Priorities for marine biomedical research

Several of the promising areas for marine biomedical research include:

1. Exploration of marine biodiversity for discovery of new pharmaceutically-active compounds

The extent of marine biodiversity remains unknown because of both the relative inaccessibility of the habitats and, in the case of microorganisms, difficulty in culturing and classifying new species. Partnerships between industry and academia need to be encouraged to provide cross-disciplinary expertise and to offset the expense of investigating the potential value of marine species in the development of new therapeutics.

2. Understanding of the molecular mechanisms for natural marine toxin action on organisms.

There are three reasons for understanding the molecular mechanism of marine algal toxins: 1) to provide structural information needed to develop drugs that will interfere with the toxin's biological activity, 2) to develop less expensive methods for detecting toxins in contaminated seafood, and 3) to use toxins as tools for investigating the biochemistry of the nervous system.

3. Development of techniques to culture species with biomedical value.

One limitation on the use of marine organisms is their availability. Harvesting of species either for research or for extraction of a biologically active compound is expensive and may deplete the natural population. Aquaculture, cell culture, microbial fermentation, and recombinant DNA techniques can provide alternative sources of material for research and drug development.

4. Expansion of drug discovery efforts beyond anti-cancer compounds

Current programs have promoted drug discovery efforts for cancer therapy through the matching of compounds from non-traditional sources (such as marine organisms) with the screening and development potential of the pharmaceutical industry. Similar approaches would benefit drug discovery efforts for other disorders including neurodegenerative, cardiovascular, and infectious diseases.

5. Encourage training and research to expand our knowledge of marine organisms.

The use of marine organisms in biomedical research and in drug development depends on knowledge of marine biology—the natural history, taxonomy, physiology, molecular biology, and biochemistry of various species. As academic biology and other science departments become increasingly subdivided into specialized fields, it is important to encourage a multidisciplinary approach to exploring the diversity of life in the ocean and to provide opportunities for students and researchers to study organisms that are representative of marine diversity.

Pursuit of the priorities outlined above will promote the development of better strategies for reducing the health problems arising from marine natural hazards, climate change, and disease-causing organisms, and will support opportunities for using marine organisms to develop new treatments for human diseases.

Introduction

When fishermen report strange episodes of memory loss associated with a bloom of the dinoflagellate *Pfiesteria* or there is an announcement of a new cancer-fighting drug isolated from a marine organism, the public becomes aware of the potential of the ocean both to threaten and benefit human health. Attention is then directed to the scientific experts to interpret these events and to make the appropriate policy recommendations. However, this requires an active research community that recognizes the links between human health and ocean processes and nurtures cooperative studies in fields as diverse as physical oceanography, public health, epidemiology, marine biology, and medicine. There is clearly a need for scientific exchange among these fields to ensure an integrated approach to issues where human health and ocean systems intersect. The workshop and this report on the Ocean's Role in Human Health were designed to initiate this type of interaction and provide guidelines for future programs.

In 1998, awareness of the dangers associated with marine driven weather events was heightened by the occurrence of an unusually strong El Niño which brought record rainfall and flooding to coastal communities. Because the majority of the world's population lives in coastal zones, the hazards associated with these events have an inordinate impact on public health. Coastal communities are especially vulnerable to storm surges, coastal flooding, and the outbreak of diseases that invade marine habitats. These natural disasters in turn stress the public health infrastructure that ensures safe drinking water, sewage treatment and disposal, and emergency medical care.

Another growing concern in the international community is the potential for global climate change due to alterations in atmospheric conditions from human activities such as the burning of fossil fuels and deforestation. Because the ocean acts as an immense reservoir for water, heat, and carbon (mineral and biogenic), an understanding of ocean processes is essential for predicting climate change. Also, because the oceanic and atmospheric systems interact dynamically, there is concern that a change in atmospheric conditions due to global warming could modify the thermal-driven circulation of the world's ocean. This dynamic circulation is essential for replenishing seawater with nutrients and oxygen and represents one of the driving forces behind changing atmospheric conditions and weather. As demonstrated during the 1997-1998 El Niño, a change in the temperature of a body of water in the equatorial Pacific can have dramatic impacts on the climates of countries as far away as Africa and Asia.

However, why should the medical community be concerned with changes in climate and weather? Experts in infectious diseases warn that in concert with changes in temperature and rainfall, there will be a change in the distribution of both waterborne and vector-borne diseases (Colwell, 1996; Haines and Parry, 1993; Rogers and Packer, 1993). Disease organisms are sensitive to changes in their environments and can quickly spread into new areas when conditions become favorable for their survival or decrease or disappear when conditions become unfavorable. Add to this the increasing mobility of human society and the probability of dispersal of disease agents worldwide increases with increasing travel and commerce.

In the past, outbreaks of cholera have been traced to shipping activities (Colwell, 1996). Current research indicates that the bacterium responsible for causing cholera, *Vibrio cholerae*, can spread through attachment to marine organisms in ship ballast water. In addition, recent studies have linked some disease outbreaks to changes in marine conditions such as prolonged periods of elevated sea surface temperature (Colwell, 1996). *Vibrio cholerae* inhabits the guts of planktonic crustaceans, specifically copepods. In this milieu, the bacteria can become concentrated into a dose sufficient to cause cholera if only a few copepods contaminate ingested shellfish or drinking water. Therefore, conditions that favor the growth and reproduction of planktonic organisms may also favor the propagation of resident cholera pathogens.

The primary source of marine-borne illness in humans is ingestion of seafood contaminated with hepatitis and caliciviruses, and pathogenic bacteria including both indigenous marine vibrios and a variety of non-marine species from human and animal fecal contamination (IOM, 1991). Traditional methods for assessing bacterial and viral contamination are not always adequate for assessing the public health risk. The increasing demand for seafood in both industrialized and developing countries, compounded by the variety of waterborne pathogens, adds to the potential for outbreaks of disease, a threat that may be offset by vigilant public health surveillance.

Other public health threats associated with marine conditions include harmful algal blooms (HABs) which have increased in frequency and geographic range in the past two decades. HABs result in human health hazards caused by dead and dying fish, adulteration of food supplies by at least five different types of toxic agents, and inhalation of these toxins when they become aerosols through wave action. Many of these toxins affect the nervous system and cause paralytic, diarrheic, neurotoxic, and amnesic shellfish poisoning, and ciguatera fish poisoning, with symptoms ranging from mild nausea to paralysis and death. As discussed *in Marine Biotoxins and Harmful Algae: A National Plan* 1992 (Anderson et al., 1993) and *The Ecology and Oceanography of Harmful Algal Blooms: A National Research Agenda* (ECOHAB, 1995), management of this problem will require research in biological oceanography to study how ocean processes affect the blooms and the distribution of different algal species, in biochemistry and molecular biology to identify algal toxins and the metabolic pathways affected by algal toxins for improved diagnosis and treatment, and in epidemiology and public health to document the association between outbreaks of illness and incidence of algal blooms, and to provide appropriate health warnings.

Scientists have investigated the algal toxins in order to understand the pathology of the illnesses listed above and to use these toxins as molecular probes of neural function. Many marine organisms have contributed to biomedicine through the unique molecules they produce. Other examples include bioluminescent and fluorescent indicator proteins, restriction enzymes used in molecular cloning, and novel antibiotics, anti-inflammatory agents, and anti-neoplastic drugs. Why do organisms produce these novel compounds that are incidentally valuable to humans? Similar to their terrestrial counterparts, marine organisms have evolved molecular strategies for evading disease and predation. Many of these strategies include the production of inhibitory chemicals that have biological activity in humans and may provide new therapies for a variety of diseases.

Finally, there has been a long tradition in the biological sciences of exploiting the properties of marine organisms as models for biomedical research. Through study of their unique adaptations to the marine environment, these organisms have contributed to our understanding of human biology from the most reductionist cellular process to the unraveling of the neural processes underlying behavior. Some of these discoveries have started new areas of investigation in cancer research, immunology, inflammatory joint diseases, kidney physiology, and neurochemistry to name a few prominent examples.

Clearly, we have benefitted greatly from our past studies of the ocean and marine life. However, challenges remain to develop more accurate forecasting of climate and weather, to assess the threat of the emergence and spread of infectious diseases, to anticipate changes in the growth and distribution of harmful algal species, and to explore marine biodiversity in order to uncover new pharmacologic agents and animal models for biomedical research.

ORGANIZATION OF THE REPORT

This report evolved from the workshop on The Ocean's Role in Human Health, hosted by the National Research Council's Ocean Studies Board. In the first part of the workshop, experts in physical oceanography, public health, infectious diseases and harmful algal blooms discussed areas in which the ocean presents risks to human health. In the second part, experts in marine pharamacology and marine biomedical models presented examples of how marine organisms have provided a source of new compounds for treating human disease and have led to discoveries of biological processes that underlie many human diseases. A complete description of the workshop appears in Appendix C, the general topics and speakers are listed below:

Part I. Hazards to Human Health From the Ocean
  • Marine Natural Disasters
      Speakers: *Claude de Ville de Goyet* (Pan American Health
                      Organization)
                  Peter Rhines (University of Washington)
                  Lynn K. Shay (University of Miami, RSMAS)
                  William Wiseman (Louisiana State University)
  • Infectious Diseases
      Speakers: *Frances Carr* (USAID)
                  D. Jay Grimes (The University of Southern Mississippi)
                  Joan Rose (University of South Florida)
                  Milan Trpis (The Johns Hopkins University)
  • Harmful Algal Blooms
      Speakers: *Lorraine C. Backer* (Centers for Disease Control and
                      Prevention)
                  Daniel G. Baden (University of Miami, NIEHS Marine and
                      Freshwater Biomedical Science Center)
                  Sherwood Hall (Food and Drug Administration)
                  Patricia A. Tester (National Marine Fisheries Service
                      NOAA)
Part II. Value of Marine Biodiversity to Biomedicine
  • Marine Natural Products
      Speakers: *William Fenical* (Scripps Institution of Oceanography)
                  Shirley A. Pomponi (Harbor Branch Oceanographic
                      Institution, Inc.)
                  Baldomera Olivera (University of Utah)
  • Marine Organisms as Models for Biomedical Research
      Speakers: *Robert Baker* (New York University Medical School)
                  David Epel (Stanford University)
                  Joan D. Ferraris (National Institutes of Health)
                  John J. Marchalonis (University of Arizona)

Following the workshop, the Committee on the Ocean's Role in Human Health met and discussed the major findings presented and outlined conclusions based on these findings. The committee structured the report using the same organization as the workshop. Hence the five sessions of the workshop correspond to Chapters 1, 2, 3, 4, and 5 of this report. Because each chapter addresses a discrete topic, conclusions are presented at the end of each chapter. Finally, this report is meant to serve as an overview, not a comprehensive discussion of all the possible connections between human health and the oceans. For this reason, health hazards such as shark bites and jellyfish stings and the secondary effects of pollution and changes in ocean-based food resources were not included in the scope of the study.

Part I

○

Hazards to Human Health From the Oceans

Part I of this report identifies areas where coordinated efforts between the oceanographic and medical communities will be required to address the risks to human health generated by the oceans and to evaluate the potential consequences of climate change for public health. There are three chapters included in Part I. Chapter 1: "Climate and Weather, Coastal Hazards, and Public Health," describes how public health is affected by marine processes such as ocean-dependent weather and climate effects, tropical storms, and estuarine and coastal circulation. Chapter 2: "Infectious Diseases," covers the various waterborne marine infectious diseases including bacterial, viral, and protozoal agents of disease. This chapter also examines the effects of weather and climate on vector-borne diseases, such as the increased prevalence of malaria, a disease carried by mosquitoes, following an El Niño event. Finally, in chapter 3: "Harmful Algal Blooms," the various syndromes resulting from exposure to algal toxins are identified and discussed with reference to the ecology and distribution of the specific algae associated with these illnesses.

There have been several recent programs that have highlighted the value of an interdisciplinary approach to the issues described in the following chapters. Brief descriptions of three of these programs appear in boxes in the appropriate chapter: HEED, in Chapter 1; the ENSO Experiment, in Chapter 2; and ECOHAB, in Chapter 3.

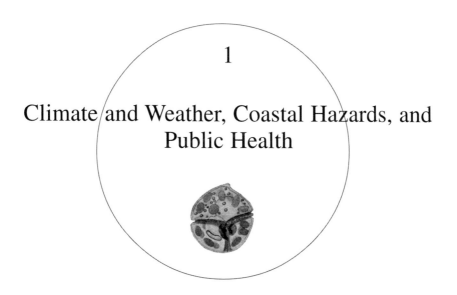

1

Climate and Weather, Coastal Hazards, and Public Health

THE PHYSICAL OCEAN ENVIRONMENT: CIRCULATION AND STRATIFICATION

Public health policymakers rely on the ocean sciences to help them develop more effective responses to marine hazardous phenomena including tropical cyclones and hurricanes, tsunamis (or long oceanic waves), toxins and pathogens in nearshore and estuarine waters, and ocean-driven weather and climate patterns. These events may either directly cause injury and death or indirectly cause the spread of various types of illness, including waterborne and vector-borne diseases, as well as illnesses associated with toxic algal blooms. The protection of public health requires a thorough understanding of the physical ocean environment for better forecasting and handling of marine disasters. This chapter will describe the major threats to public health and the ocean processes that contribute to them.

Frequently, the ability to anticipate and respond promptly to natural disasters rests on an understanding of weather systems that depend on a complex coupling of the atmosphere, land, and the ocean. The ocean absorbs immense quantities of heat, fresh water, and carbon, and hence act as the "memory" of the atmosphere and land. Although climate is seen more readily by observing the atmosphere, it is the coupled system of the ocean, the atmosphere, and the land masses that determines the evolution of climate over long periods of time. Beyond its role as a reservoir for water, heat, and carbon, the world ocean actively influences the atmosphere, as demonstrated by the El Niño phenomenon. In an El Niño, the normal upwelling of cold water along the Equator fails, and a lens of warm tropical surface water spreads across the eastern Pacific. These warm ocean waters

17

contribute moisture and energy (in the form of heat) to the atmosphere and bring unusually warm, wet weather to the west coasts of North and South America, and droughts to Australia and southeast Asia.

The ocean also serves as the medium for the culture of phytoplankton, microalgae that produce oxygen as a by-product of photosynthesis and account for about 50% of the earth's primary productivity. Through the consumption of seafood, many human populations depend on this resource. In general, the productivity of the world's oceans, which includes fisheries, is a reflection of the primary productivity of the phytoplankton. However, these phytoplankton also pose public health problems because some species produce toxins that cause various illnesses when they are consumed in seafood, as described in more detail in chapter 3. Hence, it is important to know what properties of the marine environment influence the productivity of phytoplankton.

The continued high productivity of ocean waters is dependent on the renewal of its life-giving resources: oxygen, carbon, minerals, and nutrients. This renewal occurs through the inflow of rivers and stirring of the ocean basins by great global currents that circulate waters from the surface down to the depths of all the major oceans and back to the surface again (Plate I). Without this "overturning" circulation, the surface waters would become depleted of nutrients and the deep waters would become depleted of oxygen, a stagnation observed seasonally in some freshwater lakes and in isolated marine basins and some estuaries.

At the grand scale of the world ocean, the ventilation of the deep water is achieved by cycles of evaporation and precipitation, heat exchange, winds, runoff of fresh water from land, and freezing and thawing of ice. As surface waters are cooled, they become more dense and sink, bringing oxygen-rich water to the deep. This causes stratification, where dense (cold or more saline) waters slip beneath buoyant (warm or less saline) waters. However, this atmospheric-driven sinking occurs only in very narrow, concentrated currents, whereas water rises toward the surface much more broadly. This results in the asymmetric creation of stable, stratified water masses.

Stratification is perhaps the most important physical property of the ocean for life on Earth because it determines the distribution of nutrients and oxygen. Even though the deep waters are only about 0.3% denser than surface waters, this difference is sufficient to segregate water masses and hence the sources and sinks of biological activity. Biological production depends on the input of nutrients into surface waters where there is sufficient sunlight for photosynthesis. Zones of high productivity occur where the nutrients and oxygen enriched deep waters are brought to the surface by physical forces, such as wind-driven upwelling or tides. In some areas, upwelling is driven by the force of the wind on the sea surface, pulling water up from depths of 100 m (328 feet) or more. In coastal waters and estuaries, nutrient input from rivers may be dominant. The lower density of freshwater keeps these nutrients stratified at the surface creating zones of high biological productivity.

Stratification of water masses with different temperatures and salinities also influences climate and local weather systems that in turn affect human health, both through severe storms and through changes in climate that alter the range of agents of infectious disease. This chapter will discuss why physical ocean properties have important implications for public health.

PUBLIC HEALTH PROBLEMS CAUSED BY TROPICAL STORMS AND OTHER MARINE NATURAL DISASTERS

Natural disasters involving ocean processes include phenomena such as rain, tropical storms, tsunamis, storm surges, blooms of toxic algae, pathogen contamination of coastal waters, and recurring as well as long-term climate variability. The types of problems faced by the public health system are determined both by geography and the socioeconomic status of the affected country. While the wealthier industrialized nations suffer more economic loss, poorer developing countries often face far greater loss of life, continuing incidence of disease, and longer lasting damage to social and physical structures. The potential for immediate casualties and communicable disease outbreaks often overshadows the more severe and durable long-term impacts on public health, even in developing nations. Both the direct and indirect impacts of marine natural disasters are assessed here with primary consideration given to the impacts of tropical storms, tsunamis, and storm surges. The public health issues arising from the spread of pathogens and harmful algal blooms will be discussed in Chapters 2 and 3, respectively.

Risk is a measurement of the degree of loss (human life, injuries, economic losses, etc.) expected by the occurrence of a disaster. Short- and long-term health consequences are a result of the contributions of many factors, such as:

• The type of disaster: tropical storms, storm surges, tsunamis, and gradual climate changes (from floods to drought). Loss of human lives is particularly severe following storm surges.

• The type of housing and other construction in the affected community: tropical storms will have different immediate health consequences in a poor island with highly vulnerable wooden housing than in a metropolitan area such as Miami.

• The level of economic development: although closely associated with the housing type, the economic level will determine the capacity of the community to respond and recover from the impact, therefore natural disasters affect the poor disproportionately.

• The level of preparedness of the community and health services. A well-educated population and an effective warning system will save many lives.

• The level of vulnerability as determined by local landuse practices, e.g., deforestation is believed to have contributed to the mudslides in Honduras and Nicaragua following Hurricane Mitch in 1998.

• Most critically, the vulnerability of the community as determined by the overall health and incidence of communicable diseases prior to the impact.

These factors explain why mortality and morbidity caused by the same weather event may vary widely from one country to another.

Direct Impacts on Health

Mortality, the number of deaths caused by natural disasters, is the most common indicator used by the international community to assess the severity of the health impact of a disaster. The number of lives lost provides important statistical data but can be highly misleading in determining the impact on survivors.

High mortality resulting from marine disasters is associated with tsunamis, storm surges, and flash floods resulting from tropical storms with heavy precipitation. For example, a cyclone in the Bay of Bengal in 1970 induced a storm surge causing between 250,000 and 500,000 deaths (Murty et al., 1986; Sommer and Mosley, 1973). The broad range in estimating mortality following the Bangladesh cyclone reflects the lack of reliable data on the consequences of natural disasters in most developing countries. Also, this dramatic loss of life illustrates the high-risk to the population in Bangladesh—a combination of the frequent occurrence of cyclones and storm surges with an extremely vulnerable and unprepared population. At the other end of the spectrum, accurate forecasting of tropical storms in the Caribbean region and effective evacuation policy and procedure have greatly reduced the loss of lives from recent hurricanes in the United States, although there is still potential for significant loss of life due to tropical storms (Pielke and Pielke, 1997).

The devastating 1998 tropical storm season in the Caribbean illustrates the disproportionate effects of hurricanes on developing and industrialized nations. Hurricane Georges, which struck the Caribbean and U.S. Gulf Coast in September of 1998, was responsible for an estimated 210 deaths in the Dominican Republic while causing fewer than 10 fatalities in the U.S. and Cuba (AP, 1998a).

At the end of October 1998, Hurricane Mitch brought tragedy to Central America when the storm stalled over the coast of Honduras and dropped torrential rains in the highland and coastal regions for several days. The rains caused catastrophic floods and landslides throughout the region, with Honduras suffering the heaviest losses. In Honduras, Nicaragua, Guatemala, and El Salvador there were an estimated 9000 deaths with another 9200 people reported missing. More than half of the casualties occurred in Honduras, where approximately 12,000 people were injured and 1.5 million were affected by the storm and its aftermath (USAID, 1998; UN, 1998). The devastation brought by Hurricane Mitch was not a result of poor prediction, but instead illustrates how natural disasters can result when poverty drives the development and deforestation of vulnerable areas (Copley, 1998; LaFranchi, 1998).

Tsunamis are long oceanic waves caused by earthquakes that displace the seabed. Although tsunamis occur less frequently than tropical storms, some predictions allow vulnerable communities to be warned in advance. As with a storm surge, mortality is high in low-lying coastal areas. On July 17, 1998, a shallow earthquake near the coast of Papua, New Guinea drove a wave onshore, inundating a strip of heavily populated shoreline. The first wave approached 30 feet in height and arrived 9 minutes after the earthquake (Plate II). The sudden influx of water resulted in more than 2000 deaths. Because the earthquake was so close to shore, no form of prediction and warning system could have prevented this loss of life. In many other cases, earthquake epicenters are far from vulnerable shores and warnings are effective. Elaborate warning networks are in place on many Pacific islands and around the Pacific rim.

The human health risks from the Papua, New Guinea tsunami situation were so great that officials declared a state of emergency on the Sissano coast and sealed off an area of about 120 square kilometers (45 square miles) around the lagoon. The devastation of this area forced crowding of displaced residents onto higher land. Such crowding in a wet environment, along with disruptions in the supply of potable water, favors the spread of infectious diseases such as pneumonia, cholera, and malaria. In addition, most of the injuries suffered by survivors were open wounds as a result of the physical force of the tsunami. The greatest danger to the injured therefore was infection because of limited medical care in the immediate aftermath of the disaster.

Although epidemics of waterborne (diarrheal diseases including cholera and typhoid fever) or vector-borne diseases (such as dengue fever and malaria) are a major concern, remarkably few major outbreaks have been scientifically documented in the literature following such natural disasters. Typically, in the Bay of Bengal, storms surges may cause greater problems from the salination of wells and agricultural lands than the contamination of water with pathogens. The absence of anticipated disease outbreaks may reflect several factors. First, waterborne diseases are highly preventable through public and individual environmental health measures. The very fear of devastating outbreaks is an effective incentive to improve otherwise neglected basic sanitation and water control in many countries. Second, dilution of fecal contamination by tropical storm surges in overcrowded and heavily contaminated environments may reduce outbreaks. Finally, the health indicators of the surviving population, in some instances, appears to improve in the case of a disaster such as a storm surge because the death toll is highest among the elderly, children, and the sick—groups with the greatest health problems (Chen, 1973; Sommer and Mosley, 1972).

In the case of vector-borne diseases, tropical storms, floods, and storm surges may either suppress or promote the breeding of the vector and its pathogen. Initially, the influx of water may disrupt insect vector breeding sites and decrease the rodent vector population. Later, however, breeding sites for mosquitoes (the vector for malaria, dengue, and yellow fever), while initially washed away by the floods or storm surge, increase with the use of residential water pools that build the potential for increased transmission. For example in 1963, Hurricane Flora struck Haiti shortly after the completion of insecticide spraying of the dwellings by the malaria eradication program. The proliferation of breeding sites combined with a disruption of routine control measures resulted in one of the best-documented hurricane-caused outbreaks of malaria, (Mason and Cavalie, 1965).

Indirect Impacts on Public Health

The pursuit of public health encompasses much more than the provision of medical care or the control of communicable diseases. In the constitution establishing the World Health Organization (WHO), health is defined as "a state of physical, mental, and social well-being and not only the absence of disease or infirmity" (WHO, 1946).

The delayed or indirect health impact of marine natural disasters is generally underestimated and under-reported. The long term cost to public health results from the interruption of health services, the permanent damage to infrastructure, the setback in development, and the loss of individual income. In the developing countries, electrical power and potable water shortages or rationing are daily occurrences. Following an ocean-borne disaster, the lack of electrical power (and transportation) has profound and far reaching public health consequences, affecting the operations of hospitals, water plants, and health facilities, as well as degrading the quality of the local environment. As summarized below, loss of these capabilities has the potential to affect public health more profoundly than the immediate impact of an event such as a storm surge.

Disruption of health services: Following a disaster in developing nations, health services experience a decreased ability to respond to normal demands for medical care. Hurricane Gilbert in Jamaica (1988) left a modest toll of 45 people dead, but twenty-two hospitals or health centers were out of service for an extended period of time and 90% of the hospital bed capacity was unavailable for several days to several weeks (Table 1-1; PAHO, 1988; Zeballos, 1993).

Setback in development: The economic impact of natural disasters at a national level is amplified in the health sector for several reasons. Existing resources (medicines and disposable equipment, budget, personnel) are diverted from routine medical care and disease control programs for immediate response to the perceived threats to public health. International assistance, however generous, rarely represents a significant proportion of the material emergency contribution and does not subsidize the future provision of routine care and disease control.

Loss of individual income: Even if the economy of the developing country is not significantly affected by the disaster, the most economically vulnerable popula-

TABLE 1-1   Loss of Hospital Bed Capacity

Type of Disaster	Number of Health Facilities Affected	Number and Percent (%) of Beds Lost
Hurricane Gilbert, Jamaica (1988)	22	5,065 (90%)[a]
Hurricane Hugo, Montserrat (1989)	1	67 (100%)
Tropical Storm Debbie, St. Lucia (1994)	1	25 (13%)
Hurricane Luis, Antigua (1995)	1	24 (16%)
Hurricane Luis and Marilyn, St. Kitts (1995)	1	102 (68%)
Hurricane Georges, St. Kitts (1998)	1[b]	170 (100%)

[a] Includes beds briefly unavailable in the immediate aftermath, in over 500 affected health care centers.
[b] Same hospital as in 1995.

SOURCE: Pan American Health Organization (PAHO), Regional Office of the World Health Organization (WHO), 1998.

tion is likely to suffer—poverty is the major global cause of illness and poor health (Hahn, 1996; McIntyre, 1997; PAHO, 1998a). When a family's income is reduced, there is decreased access to food, medical care, clean water, and other critical services.

Therefore, the mortality and morbidity arising from the immediate impact of marine natural disasters does not necessarily predict the long-term effects on public health. The loss of community services and secondary effects on the economy may have more serious impacts. The challenge to the international community is to help communities establish the infrastructure necessary to improve warning systems and to implement protective and preventive measures.

FORECASTING TROPICAL STORMS

Vulnerability

The vulnerability of the United States to damages from tropical storms[1] is higher now than in the past because of the growth and increased wealth of the coastal population; the population has been increasing at a rate of 4-5% per year (Sheets, 1990). Millions of people live and vacation along the coastline and are exposed to the threat of tropical storm winds, rain, storm surge, and severe

---

[1] Tropical storms is used as a general term to describe tropical storms and hurricanes, and cyclones.

weather. During this century, improved forecasts and warnings, better communications, and increased public awareness have reduced the loss of life associated with tropical storms in the United States. However, tropical storm-related damage has increased dramatically. Hurricane Andrew, in 1992, was the most costly natural disaster in U.S. history in terms of physical damage, although the loss of life was relatively low. The higher level of damages in the past decade is not from an increase in hurricane frequency, but reflects inflation, expansion of the coastal population, and the increased wealth of coastal communities (Pielke and Landsea, 1998).

The public's vulnerability is a function of the skill in forecasting the intensity of wind, rain, storm surge, and severe weather near landfall. While specific track prediction models have shown up to a 15% improvement, there has been little improvement in the prediction of intensity change (Elsberry et al., 1992). For this reason, the average length of coastline warned per storm, about 354 miles, has not changed much over the past decade. However, the average preparation costs increased six-fold in the past seven years, from $50M per storm in 1989 to an estimated $300M per storm in 1996 (OFCM, 1997). Unless the rate of forecast improvements can be accelerated, the downward trend of tropical storm casualties is not likely to continue, and the damage will continue to escalate. Track prediction is made more difficult by decadal climate variability, which leads to long-term variation in the frequency, intensity, origins, and paths of hurricanes.

Each improvement in tropical storm forecasting has been achieved by taking advantage of better observations. New strides in our abilities have always paralleled the development of new research tools; from instrumented aircraft, to radar and satellites. Because storms originate in the tropical ocean where few data are available, the scientific community has pioneered mobile observing strategies in order to provide critical observations of the storm's location and strength. One example is the production of high quality images of tropical storms produced by the SeaWiFS satellite (Plate III). These techniques have evolved to include measurements of the upper ocean and atmosphere in the vicinity of the storm. High-quality, high-resolution observations provide essential data used in determining parameters for models of atmospheric, oceanic, or coupled processes. A synergism between observations and models is required to isolate the important physical processes that will allow more accurate forecasts.

The Federal Emergency Management Agency (FEMA) requires coastal communities with limited escape routes to have completed preparation and evacuation before the arrival of gale force winds, typically 24 h to 48 h before landfall. However, an inadequate understanding of the fundamental mechanisms that inflict damage impairs our ability to provide timely warnings. Errors in wind, storm surge, and rainfall forecasts have prevented officials from accurately defining the most vulnerable regions in order to expedite required preparations well in advance of the projected landfall. Hurricane Opal provides an example of this as described below.

Hurricane Opal

During the evening of October 3, 1995, Hurricane Opal was located in the southern Gulf of Mexico [for a summary of Hurricane Opal and its impacts see the National Oceanic and Atmospheric Administration (NOAA) Service Assessment Team Report (NOAA, 1996)]. The storm had been slowly intensifying over the previous three days while drifting slowly over the Gulf of Campeche. Given the small basin size of the Gulf of Mexico, Opal could strike anywhere along the U.S. Gulf Coast within 24 h.

In the middle of the night on October 3, the storm started one of the most rapid deepening and intensification cycles that has ever been observed as it moved at 19 miles per hour (MPH) toward the U.S. Gulf Coast. Within a 5 h period, U.S. Air Force Reserve (AFRES) reconnaissance aircraft measured a steep central pressure drop (939 hPa[2] to 916 hPa), estimated surface winds increased to nearly 157 MPH, and the radius of the eye of the storm contracted from 19 to less than 10 miles. This rapid deepening presented the hurricane specialists with a major problem when the storm approached category 5 status within 12 hours of landfall without any means of alerting the public. Fortunately, over the subsequent 6 hours, the storm strength dissipated before reaching landfall later in the afternoon (Plate IV). Despite this weakening, the storm surge and wave activity were greater than anticipated and caused extensive damage along the coast, while wind damage and rainfall were less than forecast for a storm of that strength (NOAA, 1996). In forecasting this storm, why was the surge and the extent of the severe weather greater than predicted, while the wind damage and rainfall were less than predicted?

Crucial unanswered questions concerning the change in tropical storm intensity lie in three major components: (1) upper ocean heat content and the subsurface ocean structures that affect it (Elsberry et al., 1976; Black, 1983; Shay et al., 1992); (2) the inner core dynamics of storms; and (3) the winds at the jet-stream level, which are influential in steering cyclones. Important programs addressing these issues include the Tropical Cyclone initiatives of the Office of Naval Research (Elsberry, 1995) and observations and modeling by NOAA's Hurricane Research Division. These studies have demonstrated that accurate profiles of atmospheric variables obtained from aircraft deployed dropwindsondes are important for track prediction.

Ocean's Role in Modulating Intensity

The ocean's influence on tropical storm pressure and wind variations is dependent on the transfer of heat from the surface waters to the atmosphere. The recent case of Hurricane Opal demonstrated that sudden unexpected intensifica-

---

[2] hPa = hectoPascals.

tion often occurs within 24 to 48 hours of landfall, when tropical storms pass over warm, oceanic features such as the Gulf Stream, Florida Current, Loop Current or warm core rings in the western North Atlantic Ocean and the Gulf of Mexico (Plate IV). Both sea surface temperatures and the temperature regime of the oceanic boundary layer (defined as the well-mixed upper ocean layer) are needed to assess oceanic regimes where intensification is likely to occur. Sources of warm upper ocean water, carried by currents, provide a nearly continuous source of heat and moisture for moderate to fast-moving tropical storms along the lower boundary (Jacob et al., 1996; Jacob et al., 1998; Shay et al., 1992). This same effect, which occurs over the Gulf Stream, may also have led to significant increases in the surface wind field that devastated South Florida coastal communities during Hurricane Andrew in 1992 (Powell and Houston, 1996). Quantifying the effects of these oceanic features on changes in the surface pressure and wind field during tropical storm passage has far-reaching consequences not only for the research and forecasting communities, but also for the public who rely on the most advanced forecasting systems to prepare for landfall.

Storm Surges

Low atmospheric pressure causes sea level to increase underneath the storm. As the storm approaches landfall, cyclonically rotating surface winds on the right side of the eye push water onto the coast, whereas on the left side of the storm center water is driven away from the coastline. This effect, combined with tidal fluctuations in sea level and wind-generated waves, determines the storm surge or elevation of sea level. Flooding caused by the storm surge inflicts significant damage to coastal property, and frequently is responsible for the loss of life during tropical storms.

While hurricane damage in North America has primarily involved economic loss, loss of life has been the major concern with tropical storms in developing countries. As mentioned earlier in this chapter, there were between 250,000 and 500,000 immediate fatalities in the Bay of Bengal during the November 1970 storm surge, which reached 5.6 m (18.4 feet) amplitude (compounded with wind waves, the total water level exceeded 10 m (32.8 feet; Murty et al., 1986). Less easy to quantify were the loss of infrastructure, salinization of agricultural lands, and destruction of the fishing fleets (estimated at 90,000 vessels in the 1970 event). As with Atlantic hurricanes, modern satellite observations permit fairly good predictions of dangerous cyclones, but complete evacuation of the coastal region is close to impossible because most of the country is less than 33 feet above sea level. The population of Bangladesh is growing at a rate of 1.82% per year, thereby increasing the vulnerability of this nation to disastrous storm surges.

The U.S. Weather Research Program (Emanuel et al., 1995) and World Weather Research Program have identified landfalling tropical storms as a major focus of their research programs. It is in the national interest to mitigate damage that occurs after tropical storms reach landfall. Over the next decade, these issues will have a significant impact on building codes, construction technology, preparedness lead times, and evacuation procedures, all aimed at saving lives and minimizing property damage. Even if the improvement in intensity predictions is only a few percent per year, the benefit-to-cost ratio is high, leading to an improvement over the present state of forecasting.

As people alter the coastal landscape, the impacts of tropical storms will also change. For example, seawalls are built to reduce damage from storm surges or to anticipate sea level rise, but they also decrease the number of functional wetlands and the potential for migration of wetlands in the event of sea level rise. There are fewer natural areas left to sustain the coastal ecosystem, increasing the vulnerability of remaining wetlands and estuaries to the destructive effects of storms. Disruption of these habitats may have consequences for human health by affecting the distribution of disease-causing organisms such as toxic algae and by reducing the productivity of waters upon which many people rely for food. Documentation of such changes is only rarely possible; an exception being the measurements describing the impact of Hurricane Andrew on Florida and Louisiana (Stone and Finkl, 1995).

ESTUARIES AND THE COASTAL OCEAN

The role of estuarine and coastal circulation in public health is largely concerned with the transport, concentration, or dispersion of pathogenic organisms and the contribution of estuaries to marine food webs. In an estuary, water comes from two sources that both define the origin of the pathogens and contribute to the physical processes that affect the distribution of these pathogens. Pathogens derived from human activities enter the estuary through freshwater streams and rivers. Health threatening marine microorganisms (such as vibrios and toxic algae) enter with the influx of seawater from the ocean. Where fresh and salt water meet, the water column stratifies because salty water is more dense and sinks beneath the freshwater. This has an impact on the biological and physical properties of the estuary, with phytoplankton growth being assisted by strong stratification. The flow of waters in and out of estuaries is described below with reference to the effects on organisms that cause human illnesses.

To examine estuarine circulation, the mean circulation over a tidal cycle is considered with the starting assumption that tidal currents cause no net flow. If the tides are small and the river flow is great, the river water flows seaward over a wedge of salt water. Stratification is strong and dense particulate material accumulates at the toe of the salt wedge (Figure 1-1a). If tidal currents are stronger, salt is mixed upward into the upper layer and gets transported seaward. Stratification weakens, but the two-layered flow regime remains: seaward at the surface and landward below, with net transport increased in both layers (Figure 1-1b). The circulation of the estuary then greatly exceeds the river inflow. In this case,

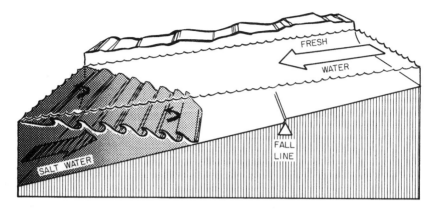

FIGURE 1-1a   Estuary circulation: if the tides are small and the river flow great, the less dense freshwater flows seaward over a wedge of salt water. Stratification is strong and dense particulate material accumulates at the toe of the salt wedge (Williams, 1962).

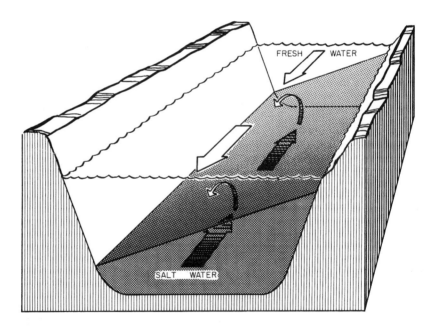

FIGURE 1-1b   Estuary circulation: if tidal currents are stronger, saltwater becomes mixed into the upper layer and gets transported seaward. Stratification weakens, but the two-layered flow regime remains: seaward at the surface and landward below, with net transport increased in both layers (Williams, 1962).

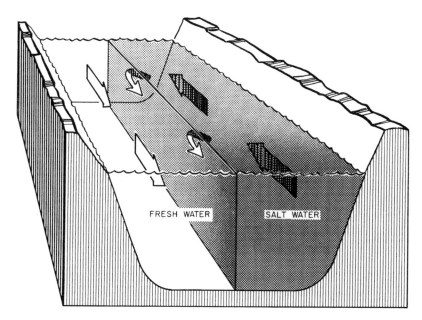

FIGURE 1-1c Estuary circulation: if tidal mixing is strong, saltwater flows in on one bank and freshwater flows out on the opposite bank with the density front visible at the surface (Williams, 1962).

---

tidal mixing is largely responsible for the quality of the water because it increases the average circulation by as much as 50 times. The rotation of the Earth causes a general displacement of the seaward flowing layer to one bank (direction dependent on location in the northern or southern hemisphere). If tidal mixing is strong, it is possible to observe inflow on one bank and outflow on the opposite bank through the density front at the surface (Figure 1-1c). These density fronts will intersect the shoreline at some point, providing a mechanism for the landward transport of particles. In fact, cysts of dinoflagellates (which can initiate a harmful algal bloom) have been observed to accumulate at the landward edge of these density fronts (Garcon et al., 1986).

Estuaries have diverse geography that affects their circulation, as does the time-variation of forcing by tides, river-flow, and solar heating. Their topography acts to weaken the fresh surface outflow, and strengthen the deep inflow, in deep channels, with the opposite bias over shoals (Valle-Levinson and Lwiza, 1995).

River inflow varies greatly in different estuaries. The ratio of estuary volume to average river inflow per day yields a rough measure of the residence time of water in the system and varies from 579 days (Puget Sound) to 307 days (Chesapeake Bay) to 49 days (San Francisco Bay). These values are subject to great seasonal variation, reflecting changes in seasonal runoff. When weak river flow rates are coupled with strong nutrient inputs, plankton blooms occur. As the plankton sink through the water column, they are consumed and decomposed, activities that consume and deplete oxygen in subsurface waters. Stratification of the water column acts to restrict mixing and thus prevents the reoxygenation of the depleted subsurface waters (Figures 1-2a and b; Table 1-2). In severe cases,

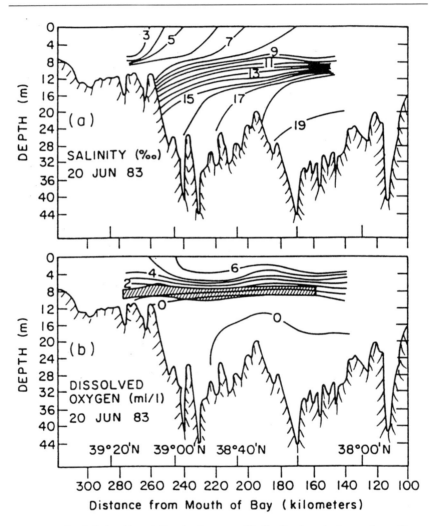

FIGURE 1-2   Salinity (a) and dissolved oxygen (b) distributions in the northern Chesapeake Bay, June 20, 1983. River water floats outward, on top of the high-salinity seawater (a). Severe hypoxia occurs below the strong salinity (and density) stratification (b). The cross-hatched region in b. shows the vertical extent of the pH minimum associated with a layer of suspended sediment (Tyler and Seliger, 1989).

TABLE 1-2   Physical Characteristics and Eutrophication Status of Selected Estuaries

Estuary	Volume (km³)	Mean Depth (m)	Mean Discharge (m³/s)	Drainage Basin Area (km²)	Eutrophication Status
Narragansett Bay	4	9.2	91	4,300	Nuisance algal blooms, summer hypoxic and anoxic events
Hudson River/ Raritan Bay	5.1	6.3	756	41,600	Nuisance algal blooms, summer hypoxic and anoxic events
Chesapeake Bay	64.5	9.4	2,430	166,800	Nuisance and toxic algal blooms, summer hypoxic and anoxic events
Mobile Bay	3.3	3	2,246	111,400	Toxic algal blooms, hypoxic and anoxic events
Mississippi River[a]	6.9	7	13,150	2,939,900	No nuisance or toxic algal blooms, no hypoxic or anoxic events
San Francisco Bay	8.4	6.3	1,966	121,900	Nuisance algal blooms, hypoxic and anoxic events

[a] Large areas of hypoxic and anoxic bottom waters, as well as toxic and nuisance algal blooms, occur in the inner shelf waters downdrift of both the Atchafalaya and Mississippi River deltas. These waters, though, formally are not considered to be an estuary.

SOURCES: NOAA, 1997a; NOAA, 1997b; NOAA, 1998a; NOAA, 1985.

anoxic conditions develop and cause massive fish kills that may present a health hazard.

Conversely, estuaries also experience river floods, occasionally flushing salt from the river entirely. In this instance, freshwater pathogens (in waste runoff) that would normally die once they encountered saltwater may survive and contaminate seafood. Flushing is also influenced by changes in stratification brought on by the spring-neap tidal cycle in some estuaries, and by wind-driven flow of nearby coastal waters. These events are stochastic, contribute extensively to the effective flushing characteristics of an estuarine system, and occur on time scales commensurate with those of plankton blooms.

Although mean conditions of flow and stratification can be monitored and modeled, this may not always help predict the transport of particulates such as dinoflagellate cysts. In some systems, it is small scale variations in estuarine circulation patterns that appear to be most important in establishing conditions conducive to harmful algal blooms. Small scale variations in flow depend on stratification, tidal currents, and topographic features that create complex circulation patterns not detected by broadly-spaced current meter arrays. Hence predicting particle transport at any particular time or location will require detailed models and/or extensive real-time monitoring systems.

Estuaries also form the conduit for the transport of high concentrations of land-derived nutrients and pollutants into coastal waters. Despite recent reductions in the input of toxic materials to U.S. waterways, concerns about coastal pollution remain regarding bioaccumulation, ecological and human health effects (NRC, 1994b). Nutrients are essential for the support of fisheries in coastal waters. As mentioned earlier, sometimes these nutrients stimulate large blooms of plankton that can result in oxygen-depleted areas, that are either hypoxic (low oxygen) or anoxic (no oxygen). This problem is particularly severe in the Gulf of Mexico, adjacent to the outflow of the Mississippi and Atchafalaya Rivers, where a low oxygen "dead zone" forms during the summer months that covers roughly 7000 square miles.

Several mechanisms have been proposed whereby nutrient-laden waters from estuaries mix across coastal waters and the continental shelf: (1) upwelling-favorable winds displace the low-salinity waters offshore, (2) winds from storms yield vertical mixing that homogenizes the water column, and (3) instabilities in flow allow the coastal current to shed eddies into the central shelf region.

In addition to transporting nutrients offshore, these processes also transport minute marine organisms such as dinoflagellates. In the Gulf of Maine, plumes of lower salinity estuarine water have been found to harbor the toxic dinoflagellate *Alexandrium tamarense* in high concentrations. Upwelling pushes the algae offshore and disperses the bloom while downwelling, when the winds reverse, causes the algae to accumulate at high concentrations along the coast where shellfish beds are more likely to become contaminated (Franks and Anderson, 1992a,b).

Similarly, coastal currents provide a potential route for the transport of toxic dinoflagellates from one region to another. The three mixing processes described above may move phytoplankton from a contaminated coast to offshore waters where currents may carry them to new downstream locations. In this new area, a relaxation of upwelling, associated with a shift in wind direction, will transport the dinoflagellates to coastal waters, hence creating conditions for a toxic bloom in a place where the phytoplankton had never been seen before. This scenario has been used to explain outbreaks of paralytic shellfish poisonings caused by a toxic dinoflagellate in the oceanic bays along the northwest coast of Spain (Fraga et al., 1988). An "upwelling index" (Bakun, 1973), based on meteorological pressure fields, has been used to investigate whether this physical feature can be used to predict blooms of toxic algae in this area. In this way, the use of hydrographic data obtained by studying physical processes may someday allow health officials to anticipate outbreaks of harmful algal blooms before the public is exposed to contaminated seafood.

CLIMATE VARIABILITY AND GLOBAL CLIMATE CHANGE

When climate change persists over long time-scales (greater than one year), the consequences for human health tend to be more serious. Regional drought, for example, can be withstood for a limited time (more so in industrialized countries) but after a prolonged period leads to famine and displacement of populations. In drought-stricken areas, higher temperatures change regional rain patterns and affect agricultural productivity, thus disrupting local food supplies. In other areas, climate variability may bring increased rainfall. Dependent on the region, these changes may alternatively increase or decrease agricultural productivity and the potential for outbreaks of waterborne diseases. Also, higher temperatures could increase or decrease the range and abundance of insect and rodent vectors of disease, possibly spreading diseases such as rift valley fever and malaria to new areas. Finally, climate change could lead to an increase in heat-related deaths, including deaths from respiratory diseases caused by air pollution which is expected to be more severe with longer, warmer summers. Although death rates increase at both temperature extremes (heat waves and extreme cold), heat-related deaths are predicted to more than offset a reduction in winter mortality (Pearce et al., 1995).

Temperature and rainfall patterns, while most often thought of as atmospheric phenomena, actually involve the interaction of the atmosphere with the ocean and the land, particularly for long-term climatic changes. In this coupled system, the ocean serves as the major reserve of heat and moisture. Attention has centered on sea-surface temperature, however, it is actually the available heat content of the upper-ocean that counts. The size of this reservoir is in part determined by salinity stratification and ice cover as well as by temperature fields. These variables, which affect the capacity of the ocean to absorb and transmit changes in atmospheric conditions, are key factors in assessing the risks posed by climate change and variability.

El Niño/Southern Oscillation (ENSO) and the North Atlantic Oscillation (NAO)

The El Niño/Southern Oscillation (ENSO) and the North Atlantic Oscillation (NAO) serve as examples of recurring weather patterns that take place on time scales longer than one year. El Niño recurs every 3-7 years, when the prevailing easterly (westward) winds of the tropical Pacific fail. This suppresses the upwelling of cold, nutrient-rich water along the central and eastern equatorial Pacific and releases a pool of warm water from the western end of the Equator. This pool of warm water propagates eastward across the equatorial Pacific towards the western hemisphere.

Sometimes the human health consequences of ENSO weather are severe. The immediate results of a warmer sea surface are increased rainfall in the eastern

Pacific and decreased rainfall in the Asian sector. Australia typically experiences severe drought. The yearly migration of the Asian monsoons also correlates with ENSO, bringing drought to some parts of Africa and India. The sea surface temperature (SST) anomalies in the central equatorial Pacific correspond to an atmospheric response that propagates along a great-circle path over North America yielding increased rain and storminess in the southwestern and southeastern U.S. Changes in temperature and rainfall due to ENSO have been postulated to lead to outbreaks of malaria and cholera, however, this proposed link is controversial. This issue is covered in greater detail in the following chapter on infectious diseases.

NAO is a mode of variability in the atmosphere over the North Atlantic. It has a very broad spectrum of time-scales, from days to centuries. It was discovered in atmospheric pressure records by 18th century missionaries in Greenland who observed that cold winters there often occurred when Scandinavian winters were mild, and vice versa. The NAO is in part a strengthening or weakening of the Icelandic low-pressure center that dominates North Atlantic weather, a statistical result of changes in the path and intensity of wintertime storms that grow over the northern Atlantic Ocean. The NAO also correlates with a large stratospheric vortex that is centered over the North Pole (Perlwitz and Graf, 1995; Thompson and Wallace, 1998).

The NAO index, which describes the waxing and waning of this phenomenon, correlates strongly with many weather variables relating to human health. Temperatures and precipitation in northern Europe, northwest Africa, and the Middle East are particularly affected. The precipitation and river-flow rate in the Tigris-Euphrates is correlated with the NAO (which corresponds to about 70% of the observed variability). Regions like this, with limited fresh-water supply, are sensitive to this degree of change. Since the 1960s the positive NAO phase has corresponded to droughts in southern Europe and the Mediterranean (Hurrell, 1995) and decreased rainfall in Morocco (Lamb and Peppler, 1987).

Global Warming, Global Change: Gradual and Abrupt Climate Change

Recent weather records reveal that over the past century the climate we have experienced is at least as warm as any century since 1400 AD, possibly due to increases in greenhouse gases. A trend towards a warmer world is emerging from the complex spectrum of natural variability (Nicholls et al., 1995). The global-average surface temperature in 1997 was the warmest of the century (Figure 1-3) and probably the warmest of the past 1000 years, while the past 8 years have included the 3 warmest years since at least AD 1400 (Mann et al., 1998). In 1998, each month set a new record for globally averaged surface temperature.

However, this change is far from uniform. A pattern of response "modes" appears to be involved, in which warming is concentrated in northern Asia (which has seen up to a 3.5 °C warming in average wintertime surface air temperatures

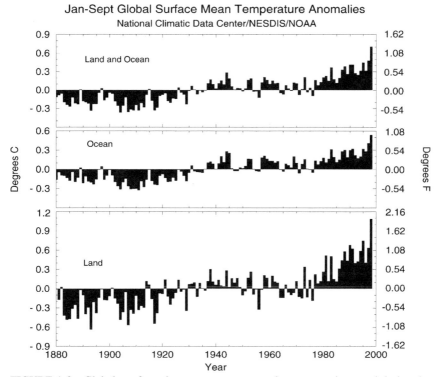

FIGURE 1-3 Global surface air mean temperatures show a warming trend during the 20th century. Since 1980 this warming has accelerated, with 1997 being the warmest year this century. Temperature changes in a complex pattern, with the oceans showing moderate increases compared with the strong warming over land, particularly at high latitudes in northern Asia and northern North America. These patterns suggest an important role for atmosphere/ocean dynamics in global warming (NOAA, 1998b).

between 1980 and 1997) with lesser warming in western North America, while large regions of the North Pacific and North Atlantic Oceans and their neighboring shores have actually cooled since the 1960s. Many climate variables are affected. Water transported in streams and rivers in North America has increased significantly, along with precipitation (Nicholls et al., 1995).

Warming of the lower atmosphere is particularly rapid over far northern land masses such as central Asia and western Canada. In Alaska, the front of the Columbia glacier has retreated 8 miles inland over the past 16 years, and no longer reaches Glacier Bay. As global surface temperatures have increased by 0.3-0.6 °C during the last century, the maximum recent warming has occurred in winter over the high mid-latitudes of the Northern Hemisphere (Nicholls et al., 1995). This warming has been especially marked for the period from 1975-1994, which correlates with unusual ENSO activity (Nicholls et al., 1995) and analysis of paleoclimate indicators suggest that 1990, 1995 and 1997 were warmer than any of the last 500 years (Mann et al., 1998). The warming trend is expected to continue, with brief warming and cooling events expected as part of the natural variability. Although there is variability in outcomes from computer models of the coupled atmosphere/ocean/land system, some predict that in a world with twice the current atmospheric CO_2 levels there will be a nearly 50% increase in high northern latitude precipitation (minus evaporation) combined with significant warming (Manabe and Stouffer, 1994). The predicted global-average warming over the next 50 years (to the year 2050) ranges from 0.7 to 2 °C, with values much greater than this over land and at high latitudes. Although such change is still very speculative, it is the common outcome of these models.

In this century, the occurrence of the two strongest El Niño events (1982/3 and 1997/8), and the occurrence of the strongest positive phase of the NAO (1972-1995) have led to speculation that global warming influences these events. This recent NAO has yielded the strongest Icelandic low observed in the past 120 years, with a northward displacement and intensification of storms and strengthened deep convection in the ocean (Dickson et al., 1996). Both ENSO and NAO affect analysis of global warming through their direct impact on surface temperatures. Hence there is an interconnection between global warming, ENSO, and NAO which complicates the prediction of future events.

Effects of global warming on other weather patterns such as tropical storms are difficult to predict. Nevertheless, recent analyses suggest that there are no global historical trends in tropical storm number, intensity or location and current thermodynamic models predict a modest (10-20%) increase in maximal potential storm intensity for a doubled CO_2 climate that is small compared with natural variations (Henderson-Sellers et al., 1998). However, the continuing rise in sea level, described below, will contribute to the impact of tropical storms through the elevation of the base for storm surges (NRC, 1998a). Potential effects of global warming on major weather events like ENSO may also influence tropical storms. ENSO shifts the regions of storm activity and frequency in the eastern and northwest Pacific and decreases the frequency of storms during the warm phase of ENSO in the North Atlantic region (Henderson-Sellers et al., 1998).

Risk of waterborne infectious diseases and vector-borne diseases also is likely to be influenced by climatic changes and ENSO events as discussed in Chapter 2 of this report. In addition, there is a risk of increased morbidity and mortality among vulnerable populations if global warming disrupts normal weather patterns and causes temperature extremes (hot or cold waves), regional flooding, and severe storms. In some areas, however, climate change may result in milder weather patterns, resulting in lower morbidity and mortality. Some of these health concerns, as well as the effects of climate change on marine ecosystems, have been addressed by the collection of marine disturbance event data in the HEED program (Box 1-1).

BOX 1-1 Health, Ecological and Economic Dimensions of Global Change Program (HEED)

In the final session of the Workshop on the Ocean's Role in Human Health, Ben Sherman and Erika Siegfried from the Harvard School of Public Health gave a presentation on the Health, Ecological and Economic Dimensions of Global Change Program (HEED). This three-year effort ending in 1998, was funded by NOAA's Global Programs and NASA and began the effort to develop a systematic methodology for collecting morbidity and mortality occurrence data across a range of marine species. The goal was to provide a baseline level of historic data on major marine ecological disturbance events to better understand and recognize changes as they occur in the world's oceans. HEED attempted to bring together the expertise of different disciplines, organize historical data in a standard format, assess the integrity and coverage of data and provide a method for standardization of data collection and analysis for the future. The resulting database was designed to be used to map reports of ecological disturbances to reveal geographic and temporal "hotspots." The occurrence of "hotspots" could serve as indicators of the impacts of climate variability such as an El Niño event. In establishing databases of this type the information entered must be independent of hypothetical outcomes. It is not possible to predict what types of queries will yield the most valuable correlations, thus it is important to assure the quality and comprehensiveness of the data collected.

The rise in sea level, estimated at 1.2- 2 cm per decade over the last century, threatens human health through the magnitude of the storm surges that, even in the past, have had a devastating effect in low lying countries like Bangladesh (NRC, 1998a). Sea level rise results from the expansion of water as it warms, the melting of polar ice, and the solid-Earth adjustment[3] (rebounding from the weight of the last glaciation). Sea level rise has already had a noticeable effect on storm-surge occurrence at low-lying coasts and islands. Because most of the non-oceanic water is stored in the polar ice sheets, changes in the stability of these sheets could have a significant impact on sea level (Nicholls et al., 1995). There has been concern that the stability of the West Antarctic Ice Sheet is vulnerable to global warming and may collapse, causing a significant effect on global sea level. It is not yet possible to predict the likelihood of collapse, but the stability of the West Antarctic Ice Sheet is currently under intensive study. During the winter of

---

[3] The upward or downward movement of the Earth's crust following the melting of the ice sheets, which "unloaded" huge weight. Glacial rebound is still being observed and has a major effect on observed sea level.

1997/98, the ice cover in the Weddell Sea was less than in any winter since satellite observations began in the early 1970s.

Abrupt change in ocean circulation has been advanced as a possible outcome of warming and increased fresh-water loading of the high latitude oceans (Broecker, 1997; Manabe and Stouffer, 1994; Rahmstorff, 1995). The great meridional overturning circulation described at the start of this chapter is thought to have been stable over the past 1000 years. Paleoclimate studies show that it was greatly altered in intensity and depth distribution during the last glacial period, and suffered frequent oscillations during other glacial periods. Computer models of the coupled ocean/atmosphere/land system suggest that we may now be entering a period of instability. Although still very speculative, such events could lead to dramatic changes in the climates of northern Europe and the Middle East over time-scales of 10 to 50 years.

The Importance of Preparedness of the Health Services

The most effective way to reduce the immediate cost in lives and human suffering from a natural disaster caused by a shift in weather patterns as occurs during an El Niño, is to improve the promptness and quality of the response of health and medical services. Coastal areas subject to surges, tropical storms, or tsunamis require a higher level of preparedness of both the emergency medical services and the health sector at large. Time and money invested on hospital contingency planning, simulation exercises, and training, not only of the first responders, but also of the entire health services, should ameliorate the outcome of ocean-driven disasters. Evidence for such progress may be seen in the recent response to the 1997/1998 El Niño. Although the actual health impact is still being evaluated by the affected countries and compiled at the international level by the Pan American Health Organization (PAHO)/World Health Organization (WHO), it is striking to note the sharp decrease in international appeals by Latin American countries compared to the 1982/83 El Niño. This suggests an improvement in the local response, a result of 20 years of increased national health preparedness and training.

Scientifically accurate and timely forecasting alone will not prevent high death tolls or decrease damages, unless the warning is transmitted, disseminated, and acted upon locally. The last decades' cyclones in the Indian subcontinent were marked by unnecessary loss of lives due to unheeded alarm. Apparently, part of the population either did not receive or fully appreciate the official warning and in other instances the design of the shelters did not accommodate the cultural norms of the community, i.e., did not include dividing walls to separate men and women (Talukder et al., 1992). Finally, even improved predictions and effective reporting are useless if no evacuation plans are available. The lower mortality associated with hurricanes on the coasts of the United States in the last 30 years is the result of several factors: improved warning systems, compliance with more stringent building codes, a successful policy of massive evacuation in the event of a serious storm, and fewer landfalling hurricanes (Pielke and Pielke, 1997). Gradual climatic changes and the possible and widely anticipated impact on disease transmission require approaches other than the traditional disaster planning aimed at improving medical readiness. Operational research on the cost effectiveness of surveillance techniques and control measures are needed to monitor and respond to disease outbreaks, natural disasters, such as floods and storms, and heat waves. Malaria, dengue, and cholera are increasing pandemics, regardless of any causal effect of the El Niño, and even the known effective control measures have yet to be implemented. Protection of the population in the United States and other developed countries in the event of climatic disturbances requires a frontal and decisive reduction of the transmission in developing countries as well. The problem is global, therefore the solution must involve the cooperation of the international community.

NEW TECHNOLOGIES FOR OCEAN ENVIRONMENTAL OBSERVATION

Physical Oceanography/Meteorology

As the problems facing life on Earth multiply, our ability to monitor them is also increasing rapidly. There has been a rapid increase in technological capacity, but a lag in implementation of these new tools. Improvements in our understanding of ocean processes will depend on a greater commitment to using new technologies to improve the baseline data used in constructing and calibrating models. Some of the current projects in ocean observations are described below.

Thanks to solid-state electronics, a large family of drifting and moored sensor packages is now at work monitoring the 3-dimensional ocean: its velocity, movement of fluid particles, temperature, salinity, and dissolved oxygen. PALACE floats, for example, now roam the ocean by the hundreds, drifting with water masses at great depth, and rising periodically to the surface to transmit their data to a satellite. The 3-dimensional ocean, with its patchy, laminated structures of biology, chemistry, and physics can now be monitored by drifting and moored sensors, and monitored from space using remote sensing. The launch of SeaWiFS in August 1997 gives us the first satellite dedicated to the global imaging of oceanic surface biological activity since the loss of the Coastal Zone Color Scanner in 1986. Comprehensive monitoring of coastal pollution and global primary productivity is now possible. Tracking of tropical cyclones and prediction of their path and intensity is vastly improved by satellite imagery, infrared images showing the sea-surface temperature, and the altimetric measurements of the Ocean Surface Topography Experiment (TOPEX)-Poseidon satellite, which measure the background surface currents of the ocean. The new network of long-range Doppler radars of the U.S. weather service aids in this process as storms approach the coast.

Integration of these new observations into computer models of the ocean and atmosphere improves both predictions and understanding of the underlying dynamics of the system. The nearly exponential increase in computing power over recent decades gives us tools that may soon resolve processes of moderate scale—like regional plankton blooms—within the context of a global model of the coupled ocean/atmosphere. However, a recent NRC report has identified a lack of available computing resources for the testing and application of climate models (NRC, 1998b).

Chemical and Biological Sensors

Biological and medical research is rapidly improving our ability to make rapid measurements of chemical and biological substances in fluids. This technology, applied to environmental measurements of the ocean, will give us high resolution 3-dimensional maps of key biological and chemical components over time. Although this technology can be prohibitively expensive, transfers of medical technologies developed for blood analysis to marine applications are feasible. In the past, only dissolved oxygen and salinity measurements have been widely collected by small electronic sensors. However, now there is active development of sensors for dissolved CO_2 (to better than 1 PPM[4]) and nutrients, which in the past required elaborate analysis of retrieved water samples. Biological monitoring with fluorometers and light-transmission sensors has been possible for some time, but new "bioprobes" for a wide range of substances are being developed aggressively. Fiber-optic chemical sensors can detect a range of variables in extremely small fluid samples (organic vapors, dissolved oxygen, CO_2, and pH; Ferguson et al., 1997; Tabacco et al., 1998). Methods for measuring specific complex biological molecules are now under development.

One of the challenges in developing new instrumentation lies in the special requirements of the ocean environment. In hospital applications, sensors for blood pH, O_2, and sugar must be replaced or recalibrated on time scales of one day or less. This would be impossible in the long term deployments typically used in oceanographic research. A discussion of sensor development, particularly those relating to oceanic carbon, is given in NRC (1993).

Collection of data from the physical, chemical, and biological monitoring systems described above creates a parallel need for comprehensive, structured databases. The efficient utilization of this information should be optimized by carefully designed, query-driven retrieval systems.

---

[4] PPM = Parts per million.

CONCLUSIONS

Marine natural disasters present challenges to both public health officials and scientists who must work together to minimize the impact of these events on human health. In some cases, this involves implementing existing technologies and preventive measures that have proven capacity to reduce human suffering. However, it is still impossible to predict and prepare for all emergencies as the recent tsunami in New Guinea and Hurricane Mitch in Central America have demonstrated. Furthermore, there are indications that the global climate is changing, affecting weather and the ocean alike. Change is nonlinear—its individual components interact in surprising ways. When human interventions occur on a back-drop of global climate change, the net effects become all the more complex and difficult to predict. The implications of the current warming trend are controversial, but this in itself argues for vigilance in monitoring changes in physical and biological systems so that predictions can be improved and problems can be identified before they become emergencies.

Potential strategies for the future include programs for public health and scientific monitoring for climate change and the health effects of climate change.

1. Support for international ocean research programs. Programs such as CLIVAR (Climate Variability and Predictability Programme) and GOOS (Global Ocean Observing System) help meet some of the needs for global monitoring. GOOS, for example, has a major initiative called "Health of the Oceans," which examines the coastal ocean and human health.

2. Closer cooperation and exchange of information across borders. Improved communication between emergency and disaster coordinators in the Western Hemisphere can help to identify common problems and solutions. In this area, the efforts of the PAHO/WHO to establish Internet links among professionals in the Americas need to be strengthened and expanded.

3. Emphasis on improving the public health infrastructure. In developing countries, storm resistant health facilities would improve medical and hospital disaster preparedness. Contingency planning and training also improve preparedness as well as support for the adoption of strict standards for hurricane and flood resistance. This emphasis on preventative measures to reduce the impacts of disasters was recognized by the Scientific and Technical Committee convened by the International Decade of Natural Disaster Reduction in June, 1998 at the World Bank in Washington, D.C. (PAHO, 1998c).

4. Establish baseline observations of the physical ocean and its ecosystems to monitor global change. Change can only be evaluated in the context of past experience. Therefore, our ability to understand future events will depend on the quality of the observations and analyses of that are collected now. This will require the implementation of newly developed technologies and the establishment of databases to design and evaluate models.

The next decade will be a crucial period for monitoring climate. The uncertain predictions of climate models will be tested and the changes in the environment may be more striking and heterogeneous than currently predicted. Achievement of the above recommendations will require a cooperative effort among the social, political, and economic sectors of the community, both local and internationally; potentially through close coordination between the scientific community and various local and international agencies such as the U.S. Agency for International Development and PAHO/WHO.

2

Infectious Diseases

WATERBORNE DISEASES

Introduction

Societies depend heavily on the ocean for food, transportation, recreation, waste disposal, and minerals. Clearly, some of these uses are in conflict, and four of these five major uses provide a vehicle for transmission of disease agents to humans. The primary source of marine-borne illness is seafood (Czachor, 1992; Lipp and Rose, 1997). Currently seafood is in high demand, both in industrialized and emerging nations, and it continues to be a source of infectious diseases in humans throughout the world (Garrett et al., 1997; IOM, 1991). In the United States, the demand for seafood is so high that fish and fish products were the third leading contributor to the U.S. international trade deficit, reaching a staggering $4.2 billion in 1997 (McCarthy, 1996; NMFS, 1998).

Transportation has also contributed to waterborne infectious diseases through the consumption of contaminated water and/or seafood by ship passengers and crew. There is also evidence that ships are responsible for the dissemination of exotic species, including human pathogens, through the discharge of bilge water into coastal waters (NRC, 1996). Recreation is another source of exposure through ingestion of seawater or contact of skin and/or mucous membranes with seawater. Finally, disposal of human wastes in the ocean can lead to diseases of plants and animals, including humans.

In the final decade of the 20th century it appears that waterborne diseases of humans are as prevalent as they were at the start of century. For example, waterborne illnesses continue to be a major killer of children throughout the world

(ICDDR,B, 1998). It is estimated that 60% of the world's population lives in coastal zones, the area within a few kilometers of the shoreline. Disease incidence is increasing worldwide, promoted by both natural phenomena such as El Niño and human activities, including sewage disposal. Ancient diseases like cholera still cause epidemics (Plate V) while new agents of disease (e.g., hepatitis E and *Vibrio vulnificus*) continue to be discovered. Although the pathogenesis of diseases such as cholera and dysentery is well understood, the cause of outbreaks of these diseases is unresolved. Experts still debate how cholera epidemics arise despite a detailed understanding of the genetics, chemical structure, and mode of action of the cholera toxin. Cholera is known to reside in human hosts and spread by the oral fecal route, however, epidemics may be seeded by vibrios that reside in estuaries and other saline waters and infect people through contaminated drinking water and seafood. The molecular structure of recently discovered disease agents has confirmed their existence on Earth for thousands of years. The apparent emergence of these pathogens could be the result of anthropogenic influences or may reflect more sensitive modern detection technologies. Whatever the cause, it is clear that infectious diseases, including waterborne diseases conveyed by the ocean, still plague humankind.

The Agents of Waterborne Disease

The principal agents of diseases that derive from seawater and seafood are viruses and bacteria. Most disease appears to result from ingestion of contaminated seafood (IOM, 1991), although accidental ingestion of seawater (e.g., recreational exposure or contamination of potable water with seawater) is another important route of infection. Some agents enter the human body through wounds (e.g., puncture wounds from sea urchins and cuts from fishing gear) others—*Leptospira*—although rarely contracted from seawater, enter through broken skin, or penetrate mucous membranes, and a few—avian schistosomes—penetrate unbroken skin.

Viruses are obligate parasites and require living cells for reproduction. Certain viruses, however, can survive in seawater for long periods of time (e.g., hepatitis A, poliovirus) and are concentrated by marine bivalve mollusks, such as oysters and clams. Human waterborne viral infections result from contamination of seawater or seafood by sewage.

Unlike viruses, most of the pathogenic bacteria do not require human hosts for replication. Indeed, some are naturally occurring in estuaries and the coastal ocean, and can grow on, or within, many animals, ranging from zooplankton to fish, and on marine plants, ranging from phytoplankton to macrophytes. In general, bacteria that cause seawater- and seafood-borne diseases in humans come from two different sources. The foreign or **allochthonous bacteria** come from humans and other animals by means of fecal contamination: sewage outfalls, septic tanks, and land surface runoff. One exception to this generalization for allochthonous bacteria is *Leptospira interrogans*; it lives in the urinary tract of mammals, especially rodents, and enters water by means of contaminated urine. The indigenous or **autochthonous bacteria** are usually commensals of marine plants and animals, although some may be free-living in the water. For example, the vibrios are indigenous to estuaries and the ocean, and several species are human pathogens. *Vibrio cholerae* causes human cholera and *V. parahaemolyticus* causes gastroenteritis and wound infections. Because of the popularity of seafood in Japan, *V. parahaemolyticus* is one of the most common causes of gastroenteritis there (IASR, 1996). Examples of the more common microorganisms that cause human disease and are conveyed by seawater are listed in Table 2-1. Detailed discussion of these common microbial agents of disease can be found in most textbooks of microbiology, including Howard et al. (1994). Grimes (1991) has reviewed the literature on estuarine bacteria capable of causing human disease.

TABLE 2-1 Selected List of Major Agents of Waterborne Disease Conveyed by the Coastal Ocean and Their Usual Routes of Transmission to Humans

Agent	Disease	Usual Transmission Route
Viruses		
Hepatitis A virus	Infectious hepatitis	Seafood[b], water[c]
Hepatitis E virus	Hepatitis	Water[c]
Caliciviruses	Gastroenteritis	Seafood[b], water[c]
Rotaviruses	Infantile gastroenteritis	Water[c]
Astroviruses	Gastroenteritis	Seafood[b], water[c]
Enteroviruses	Varied	Water[b]
Autochthonous Bacteria[a]		
Mycobacterium marinum	Granuloma	Water[d]
Vibrio alginolyticus	Wound infections	Water[d]
Vibrio cholerae	Cholera	Seafood[b], water[c]
Vibrio parahaemolyticus	Gastroenteritis, wound inf.	Seafood[b], water[d]
Allochthonous Bacteria[a]		
Escherichia coli	Dysentery, gastroenteritis	Water[d]
Leptospira interrogans	Leptospirosis	Water[d]
Listeria monocytogenes	Listeriosis	Seafood[b]
Morganella morganii	Scromboid food poisoning	Seafood
Salmonella species	Typhoid, gastroenteritis	Water[c]
Shigella species	Bacillary dysentery	Water[c]
Nematodes		
Anisakis simplex	Anisakiasis	Seafood[b]

[a] Autochthonous = indigenous to the system; allochthonous = foreign to the system
[b] Raw or undercooked seafood
[c] Water ingestion—accidental during recreation or potable water contaminated with seawater or feces
[d] Water contact—usually accidental, during recreation or occupational exposure

In addition to the viruses and bacteria, there are a few waterborne animal parasites and fungi that cause human disease. The best known are the avian schistosomes that cause swimmer's itch, the nematode that causes anisakiasis (Deardorff and Overstreet, 1991), and the yeast that causes candidiasis. Outbreaks of infection by these agents are relatively rare (toxic algae are also microbial agents of human disease, and will be discussed in Chapter 3 of this report). Stings inflicted by anemones, jellys, and stingrays associated with recreational use of coastal waters provides sites for infection. Such infections usually result from autochthonous bacteria such as the vibrios (Thomas and Scott, 1997).

Detection and Prevention

Classic detection methods for microbial agents of disease have relied on microscopy and cultivation. Microscopy is useful only for the larger organisms that can be identified by their morphology. Fluorescence microscopy increases the options for visual identification by using fluorescent dyes coupled to bacteria- or virus-specific antibodies or gene probes. The traditional method for detecting viruses and bacteria is cultivation. Suspect samples of water and seafood are inoculated into nutritive broths or on nutritive solid media (bacteria), or tissue cultures (viruses). Positive cultures allow for the identification of the agents. In some tests, samples are injected into animals, which are then monitored for symptoms of disease.

Although cultivation is often successful in isolating pathogens, it is now recognized that many bacteria remain undetected because they are in a viable but nonculturable state (Grimes et al., 1986; Roszak and Colwell, 1987; Xu et al., 1982). They can be visualized by microscopy, but they cannot be grown using currently available media and protocols. Most of the bacteria listed in Table 2-1 are capable of entering a nonculturable phase (Grimes et al., 1986; Roszak and Colwell, 1987), and no methods for their cultivation from this phase have yet been formulated. Hence, testing water and seafood for the presence of pathogenic bacteria by using culture techniques can be misleading.

Detection and counting of specific bacteria, historically referred to as fecal indicator bacteria, have comprised the generally accepted method for estimating the public health safety of both drinking and recreational water and seafood (Dufour, 1984). This technique relies on the culture of coliforms, a term used to include several genera and species, including *Escherichia coli, Enterobacter aerogenes, Klebsiella pneumoniae,* and even some *Salmonella* serotypes, thought to derive from the colon of warm blooded animals. Originally, this group of bacteria was employed to estimate the extent of fecal contamination of water samples and, hence, the potential of enteric disease (Dufour, 1984). However, it was realized as early as 1900 that some coliforms are frequently associated with natural bodies of water including estuaries. *Klebsiella pneumoniae,* for example, is commonly associated with plants and can wash into estuaries and the coastal ocean during heavy rains. If present, these non-sewage-related coliforms will cause falsely high estimates of pollution, making the use of total coliforms as an indicator of enteric disease questionable and sometimes misleading. Although the introduction of the fecal coliform count improved the specificity of the test, false positives from other non-fecal bacteria remain a problem. Today, it is accepted that the fecal coliforms provide public health practitioners with a useful indicator; however, it is also widely recognized that the fecal coliform test has some serious disadvantages:

- Not only fecal, but also non-fecal bacteria such as *K. pneumoniae*, make up the fecal coliform group, reducing the accuracy of the test;
- Fecal coliforms have little, if any, quantifiable association with specific pathogens that are important in human disease, including viral diseases (Table 2-1);
- Fecal coliforms survive for a long time in aquatic habitats, notably in estuaries and in shellfish, either in the detectable (culturable) or dormant (nonculturable) state; and
- Fecal coliforms do not provide a meaningful indication of the disinfection of water, wastewater, and seafood because commonly employed procedures (e.g., chlorination and ultraviolet light) accelerate the transition of these bacteria from a culturable to a nonculturable state.

In the early 1980s, the U.S. Environmental Protection Agency (EPA) recommended replacing the fecal coliform recreational water index with *Escherichia coli* and/or enterococci counts in freshwater and enterococci counts in marine and estuarine waters. In one epidemiologic study (Cabelli et al., 1982) the incidence of enteric disease among swimmers was compared with incidence of selected fecal indicator bacteria including coliforms, fecal coliforms, *Escherichia coli*, and enterococci in the water off bathing beaches. There was a high correlation between the incidence of enteric disease and the levels of enterococci found in marine and estuarine waters. Even so, the only coastal states that have adopted the EPA recommended standards for marine and estuarine waters include Connecticut, Hawaii, Maine, New Hampshire, and South Carolina (NRDC, 1998).

Finally, it should be noted that certain strains of *E. coli* are themselves pathogenic for humans (e.g., *E. coli* 0157:H7), and these toxin-producing strains can survive in seawater in the viable but nonculturable state for several days to weeks (Grimes and Colwell, 1986). For example, a recreational outbreak of *E. coli* disease occurred in a water park in June, 1998 (CDC, 1998a), infecting 26 children and causing one death. Clearly, *E. coli* is more than an indicator of fecal pollution; it may be a preferred indicator of contamination for freshwater and possibly low salinity water. Certain pathogenic strains of *E. coli* are themselves candidates for surveillance.

The fecal coliform test continues to be the standard used for shellfish and waters where shellfish are grown, as established by the National Shellfish Sanitation Program (NSSP; FDA, 1996). Although numerous studies have reported that fecal coliforms do not correlate with any of the current diseases that can be contracted from eating raw or undercooked oysters (Goyal et al., 1979; Koh et al., 1994), there is a reluctance for the NSSP to abandon them as its index of public health safety.

Some investigators have suggested that enterococci may be the preferred indicator, but these microorganisms have some of the same limitations as the fecal coliforms (Koh et al., 1994). Because of these limitations and reliance on the existing database for fecal coliforms, the enterococci have not been universally accepted by state, county, and local regulatory agencies as a substitute indicator group, even though some states have adopted the new *E. coli* and enterococcus standards. In addition, the presence of enterococci or *E. coli* does not always correlate with the presence of viruses.

New indicator systems are being proposed, some of which are based on molecular genetic methods. The coliphages, viruses attacking the coliform bacteria, have been suggested as an indicator of risk of infection with human enteric viruses, including hepatitis A virus. They have a quantifiable association with specific viral pathogens representing an improvement over fecal coliforms (Paul et al., 1997). Presence of *Clostridium perfringens* spores may be helpful in distinguishing long-term accumulation or movement of sewage in aquatic habitats. This is because *C. perfringens* is fecal specific and the spores can survive for many years (Emerson and Cabelli, 1982). Of greatest interest, however, are the gene probes now being used in clinical laboratories to detect pathogenic microorganisms. Gene probes can be highly specific, capable of detecting genetic sequences common to, or evolutionarily conserved in, pathogens of concern. For example, gene probes can be used in environmental testing to detect, directly identify, and quantitate pathogens such as *Salmonella* thus obviating the need for culturing (Brasher et al., 1998). Also, amplified nucleic acid sequences, using the polymerase chain reaction (PCR), allows detection of very small numbers of bacteria and viruses in a given sample. For example, Brasher et al. (1998) have developed a series of gene probes that show promise for the direct detection of enteric pathogens in seawater and seafood. Monoclonal antibodies also have been shown to be useful for direct detection of pathogens in estuarine and seawater samples, especially when coupled with fluorescent dyes (Huq et al., 1990). Finally, the use of immunoglobulin A as a human-specific indicator of fecal pollution has received some recent interest (Middlebrooks, 1993). It is predicted that the new biotechnological methods will eventually replace the cumbersome and inaccurate fecal indicator tests for public health assessments of the safety of water and shellfish. For a recent review of detecting viruses in the environment, the reader is referred to Metcalf et al. (1995).

Coastal Oceans, Estuaries, and Waterborne Disease Transmission

Allochthonous materials enter estuaries and the coastal ocean by four routes: point sources, nonpoint sources, direct disposal or discharge, and atmospheric fallout or rainout. Point sources include sewage outfalls and industrial wastewater outfalls. Nonpoint sources include storm runoffs, storm sewer outfalls, and rivers. Both point and nonpoint sources of pollution have the potential to contaminate receiving waters with human pathogens. In past years, disposal of sewage sludge contributed pathogens to both the water column and marine sediments. Direct disposal of wastes into the ocean is now largely prohibited as a result of the London Dumping Treaty (Zeppetello, 1985) and the Ocean Dumping Act of 1988, and, therefore, refuse discharge is no longer a major source of human pathogens in the ocean. A recently recognized source of human pathogens is bilge and ship ballast discharge. Ships that take on ballast water in foreign ports can inadvertently take on water contaminated with pathogens and then discharge those pathogens into previously uncontaminated waters. The National Research Council recently published a report on the introduction of non-native species by ballast water (NRC, 1996), and McCarthy and her colleagues have studied this problem in the Gulf of Mexico (McCarthy and Gaines, 1992; McCarthy and Khambaty, 1994; McCarthy, 1996). The scope of this problem is not yet clear, but ship ballast discharge has the potential to spread microorganisms throughout the world's coastal ocean. Less clear are contamination problems from atmospheric fallout or rainout, but a recent publication suggests that pathogens can be spread by air (Rosas et al., 1997). Contributions due to atmospheric fallout or rainout occur, but are not well understood (Rosas et al., 1997).

Point sources of allochthonous materials continue to be an important source of pathogens in the estuaries and the coastal ocean of the United States. Sewage treatment plants in the United States discharge approximately 2.3 trillion gallons of effluent into marine waters annually, and more than 2.8 billion gallons of industrial wastewater are released daily (NOAA, 1998c). Even though much of this effluent is receiving some type of treatment prior to discharge, it is still contributing pathogens, nutrients, and toxic chemicals to the coastal ocean of the United States. While point sources therefore continue to cause degradation of coastal areas, it is now generally believed that nonpoint sources are equally if not more important. Nationwide, it has been estimated that indirect loadings account for more than half of the suspended solids, nutrients, fecal coliforms, and metals entering coastal waters annually (NOAA, 1998a). The two largest sources of nonpoint pollutants are agricultural runoff and storm sewer discharge.

It is estimated that over 9.1 billion gallons of domestic sewage and industrial wastewater enter the coastal ocean of the United States each day (NOAA, 1998a). When combined with accidental spills (e.g., oil spills) and nonpoint source runoff, the net result is a vast and complex mixture of chemicals, both organic and inorganic, entering the coastal ocean of the United States. These chemicals primarily cause two alterations in coastal ocean ecosystems. Some serve as growth stimulating nutrients, either directly as sources of food for marine plants, animals, and microorganisms or indirectly as essential growth factors (e.g., vitamins). Other chemicals are toxic to living organisms, causing alterations in growth (e.g., stunting and neoplasms) or death. In some cases, chemicals that are toxic to some forms of life can actually serve as growth stimulating nutrients for others, especially microorganisms. A good example of this is the ability of carcinogenic polycyclic aromatic hydrocarbons (PAHs) to serve as a growth medium for certain marine vibrios, including pathogenic vibrios (West et al., 1984). Another example is the shift in bacterial communities that was described in the mid-Atlantic Ocean by Grimes et al. (1984). Thus, it is possible that certain toxic chemicals entering the coastal ocean of the world have multiple effects. Some of these chemicals could be making fish and marine mammals more susceptible to disease by altering their resistance (immune system), while at the same time stimulating the growth of agents capable of causing the disease. It is worth noting that the vibrios are capable of degrading a wide variety of organic compounds, including aliphatic and aromatic hydrocarbons, and this generalization includes the pathogenic vibrios. However, proper experiments have not been done to establish linkages between toxic chemicals, declining fisheries, human disease, and densities of pathogenic vibrios.

Most of the waterborne and seafood-borne diseases throughout the world are caused by viruses. The major agents of waterborne viral disease are listed in Table 2-1, and they include several well-known groups. The agents in these groups are all RNA (Ribonucleic Acid) viruses, and most have not been cultured. Viral hepatitis is now known to be caused by two related viruses, hepatitis A and E. These viruses are very resistant to environmental extremes, including toxic chemicals, and they are concentrated from water by filter-feeding shellfish, especially oysters. The human caliciviruses include the Norwalk virus and all cause gastroenteritis. It is estimated that most shellfish-borne and waterborne disease in the United States is caused by the Norwalk virus (GAO, 1984). None of the caliciviruses can yet be cultured and they are detected by antibody-based tests (e.g., Enzyme-Linked Immunosorbent Assay [ELISA]) and RT-PCR (reverse transcriptase PCR). The NRC Committee on Evaluation of the Safety of Fishery Products reviewed the incidence of Norwalk and Norwalk-like agents (IOM, 1991). The rotaviruses and astroviruses are responsible for widespread diarrheal disease, especially in children. It is estimated that rotaviruses kill 870,000 children in the world annually (U.S. Department of Health and Human Services, 1998). As described for the caliciviruses, their detection is dependent on RT-PCR assays and immunologic procedures. Recently, a rotavirus immunization procedure was approved for children. Finally, the enteroviruses represent a large and diverse group of viruses that include the poliovirus. Enteroviruses cause a variety of clinical syndromes, including aseptic meningitis, paralysis, myocarditis, rash, pneumonia, fever, and undifferentiated febrile illness. The only common link among these viruses is their mode of transmission (fecal-oral route, which can include water and food as intermediaries) and their nucleic acid content is RNA. It is doubtful that global climate change events will have any significant direct effect on these viruses, because they are not conveyed by vectors and are incapable

of growing in ocean habitats. However, changing demographics and anthropogenic pressures (e.g., sewage contamination) very definitely contribute to the epidemiology of these diseases. Avian schistosomes are a problem for recreational
users of coastal waters, causing what is known as swimmer's itch or clam digger's
itch. The cercariae (free-swimming larvae) of these schistosomes penetrate human skin; but humans are not hosts for the parasite and the cercariae die in the
skin causing mild to severe itching due to an allergic reaction to the foreign protein. Human anisakiasis is caused by a nematode that infects humans via raw or
inadequately prepared seafood (Deardorff and Overstreet, 1991). The incidence
of this parasitic disease is very low in the United States, but may be on the increase
because of the current popularity of raw fish as a delicacy (e.g., sushi and sashimi).
In Japan, over 3,000 cases of anisakiasis are reported annually (IOM, 1991). It is
difficult to predict how global climate change events would affect these two parasites. It is possible that increased water temperatures could disrupt life cycles of
these parasites by changing the range of their hosts (e.g., snails, birds, fish, marine
mammals) and, hence, accessibility to humans. Demographics (e.g., continued
population shifts to the coast) and cultural food preferences (e.g., consumption of
raw fish) could also have a definite effect.

Global Climate Change and Infectious Disease

One of the first documented associations of a human pathogen with an estuarine animal subject to climate-induced fluctuations was the *Vibrio parahaemolyticus*–zooplankton relationship described by Kaneko and Colwell (1973) in the
Chesapeake Bay. They determined that *V. parahaemolyticus* overwintered in
Chesapeake Bay sediments and entered the water column when water temperatures reached 14 ± 1 °C. Upon entering the water column, the vibrios became
associated with zooplankton, primarily on the surfaces of copepods. The abundance of the vibrios increased in direct proportion to the growth of the copepod
population. Kaneko and Colwell (1973) also demonstrated a direct relationship
between water temperature, numbers of zooplankton, and *V. parahaemolyticus*.
Over 10 years later, Watkins and Cabelli (1985) observed a similar relationship
between *V. parahaemolyticus* and zooplankton in Narragansett Bay. In addition,
they demonstrated that fecal pollution had an indirect effect on *V. parahaemolyticus* densities. Nutrients associated with the wastewater stimulated the growth
of phytoplankton boosting zooplankton populations, which in turn supported
greater densities of the vibrios. Clearly, coastal ocean nutrient enrichment can
increase the prevalence of pathogenic vibrios, indicating that climatic events that
change the abundance of plankton also have the potential to affect the spread of
disease. It is interesting to note that elevated water temperatures have been impli-
cated in the largest reported outbreak in North America of shellfish-borne *V.
parahaemolyticus*. During July and August, 1997, 209 persons became ill and
one person died from consuming raw oysters harvested from California, Oregon,

and Washington in the United States and British Columbia in Canada. All 209 infections were culture-confirmed as *V. parahaemolyticus* (CDC, 1998b). In early summer 1998, the Food and Drug Administration warned consumers not to eat raw oysters from Galveston Bay, Texas, because they might contain harmful levels of *V. parahaemolyticus* (FDA, 1998). By the end of July, 368 persons had become ill from consuming raw oysters harvested from Galveston Bay, Texas; so far *V. parahaemolyticus* has been confirmed in 66 of those cases (AP, 1998b).

It has become clear that *Vibrio vulnificus* also has a preference for warmer temperatures, making it a candidate for the list of pathogens that could be influenced by climate-induced increases in water temperature. Motes et al. (1998) investigated the temperature and salinity parameters of waters where oysters were linked to seafood-borne *V. vulnificus* infections, and found that abundance of this pathogen was directly correlated with water temperature. Numbers of *V. vulnificus* increased with water temperatures up to 26 °C and were constant at higher temperatures. Salinity also played a factor and high *V. vulnificus* levels were associated with intermediate salinities (5 to 25 ppt[1]).

Colwell (1996) noted that coincident with an outbreak of cholera in Peru, an El Niño event that began in 1990 had warmed the nutrient- and phytoplankton-rich waters off the Peruvian coast. The Peruvian outbreak, which began in January 1991, quickly spread to most of the neighboring countries in South America (Mata, 1994). In 3 weeks, the epidemic had covered over 2,000 km of coastal Peru and caused 30,000 cases of cholera; it claimed 114 lives in the first 7 days. Since *V. cholerae*, like *V. parahaemolyticus*, associates with planktonic copepods (Colwell, 1996), it can be hypothesized that the South American cholera outbreak resulted from increased levels of *V. cholerae* that grew in response to the warmer water and nutrients in terrestrial runoff from increased rainfall. Although this conclusion was not reached by direct experimentation, the *V. parahaemolyticus* model indicates that this proposed link is worthy of further scientific investigation.

More recently, Colwell noted a relationship between sea surface temperature in the Bay of Bengal and cholera cases reported in Bangladesh (Colwell, 1996). A plot of the percent of diarrheal disease caused by *V. cholerae* versus seasonal sea surface temperature data obtained by remote sensing, revealed a correlation between the cholera case data and increasing water temperature. As with the Peruvian hypothesis, direct experiments to relate the two observations were not performed. However, the data are compatible with the ecology of the vibrios, and should be followed up with carefully planned interdisciplinary studies including epidemiology, meteorology, and oceanography. Plate VI shows that sea surface temperature was elevated from December through April off the western coast of South America during the 1997-98 El Niño. Sea surface height (Plate VII), how-

---

[1] ppt = parts per thousand.

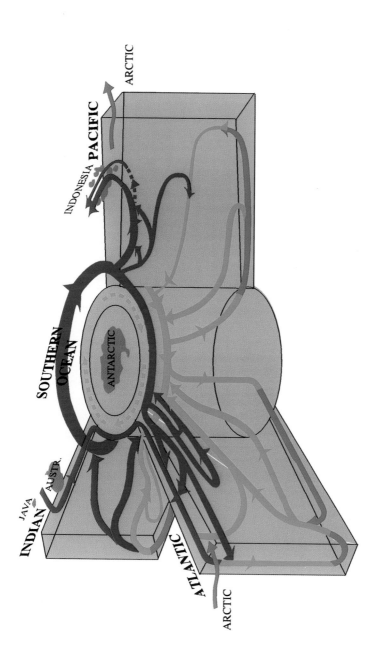

Plate I: Arteries of the ocean circulation carry warm water to the North Atlantic where it is cooled by the Arctic cold air masses. This cooling makes the water denser and it sinks to the bottom, forming a southward-moving water mass that flows around Antarctica, then filling the world ocean basins and gradually returning to the surface. Nutrients brought up to the sunlit surface layers can then support the growth of plankton (after Schmitz, 1996)

ne = 57 minutes

Plate II: Snapshot from a preliminary simulation of the 1998 New Guinea tsunami illustrating the concentrated surge as the wave hit the coastline (USGS, 1998).

24 August 1998 17:09 UTC SeaWiFS Project/NASA/GSFC and ORBIMAGE

Plate III: Satellite image of Hurricane Bonnie off the coast of Florida on August 24, 1998. Image taken by NASA/GSFC SeaWIFS satellite. Hurricanes are fueled by the warm tropical ocean, and are sensitive to ocean temperatures along their paths.

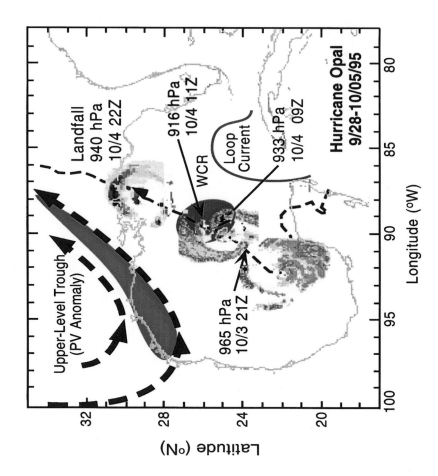

Plate IV: Storm track of Hurricane Opal in the Gulf of Mexico showing the pressure drop as the storm passed over the Loop Current warm core ring (red, WCR). The atmospheric upper-level trough (blue) influenced the steering of the hurricane as it approached landfall. The diagram is based upon TOPEX altimetry data and post-storm AVHRR images. (Adapted from Marks and Shay, 1998; Shay et. al., 1998)

Plate V: Patients suffering from cholera in a Bangladesh hospital. Photo courtesy of D.J. Grimes.

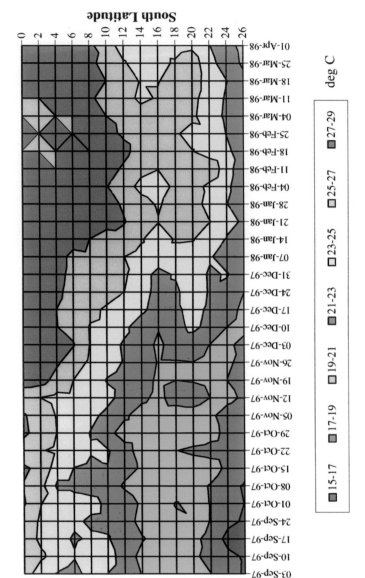

Plate VI: Sea Surface Temperatures (SST) data near the western coast of South America show the temperature started rising quickly in November 1997, and remained high throughout the spring of 1998. This chart was generated as part of an EPA funded project: "Global Climate Change and Infectious Disease: Application of Remote Sensing in Cholera Prediction," involving R. Colwell, A. Huq, J. Patz, A. Gil, B. Sack, B. Lobitz, and B. Wood. SST data source: JPL Physical Oceanography DAAC AVHRR Multi-channel SST.

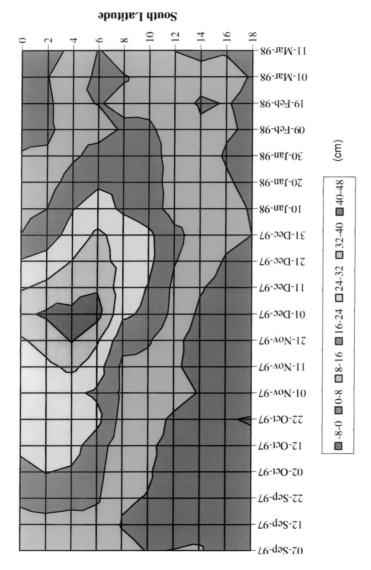

Plate VII: While Plate VI showed the warm water persisting through the spring of 1998, the elevation of this water mass (Sea Surface Height, SSH) off the coast of Equador (4°S latitude) peaked in December 1997, true to its namesake, "El Niño." This chart was generated as part of an EPA funded project: "Global Climate Change and Infectious Disease: Application of Remote Sensing in Cholera Prediction," involving R. Colwell, A. Huq, J. Patz, A. Gil, B. Sack, B. Lobitz, and B. Wood. SSH data source: University of Texas TOPEX Sea Surface Anomalies.

Plate VIII: Reverse colored (warmest is deepest blue; coldest is red) sea surface temperature image that shows the strong shoreward intrusion of Gulf Stream water (darkest blue, 28 °C) into the nearshore regions of the North Carolina coast. The Gulf Stream and meanders of Gulf Stream water serve as a transport mechanism for *Gymnodinium breve* red tide cells onto the continental shelf in the U.S. South Atlantic Bight. Image from the NOAA-9 polar orbiting satellite (AVHRR-advanced very high resolution radiometer) on October 31, 1987; image provided by Tom Leming, National Marine Fisheries Service (NMFS), NSTL, MS.

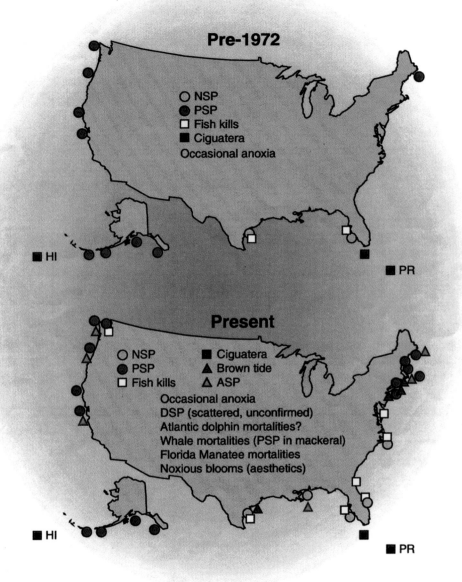

Plate IX: These maps depict the HAB outbreaks known before (top) and after (bottom) 1972. This is not meant to be an exhaustive compilation of all events, but rather an indication of major or recurrent HAB episodes. Neurotoxic shellfish poisoning = NSP, paralytic shellfish poisoning = PSP, and amnesic shellfish poisoning = ASP (Anderson, 1995).

Plate X: In densely populated habitats, marine plants and animals produce chemicals to protect them from predation and overgrowth. Some of these bioactive chemicals have potential value as pharmaceuticals. Photo courtesy of William Fenical, Scripps Institution of Oceanography.

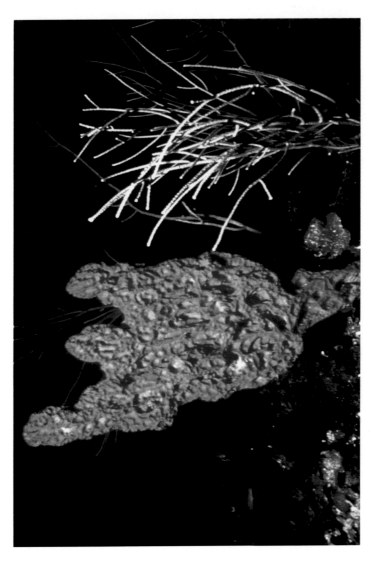

Plate XI: Sponges are dominant components of many marine ecosystems and provide a source of unique chemicals with pharmaceutical potential. This bright orange sponge is called *Teichaxinella morchella* and was photographed in the Bahamas at a depth of 100 feet on a deep water reef. This species has several interesting bioactive compounds, one of which has antitumor activity. Photo by John K. Reed, Harbor Branch Oceanographic Institution.

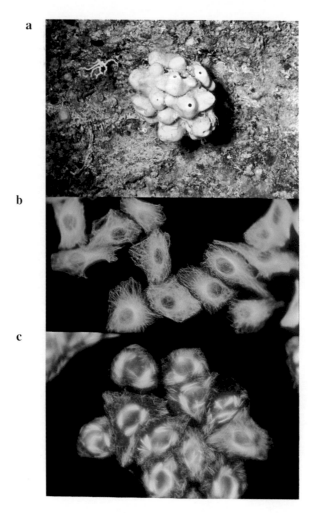

a

b

c

Plate XIIa: The deep-water marine sponge, *Discodermia dissoluta*, from which the compound discodermolide is obtained. This sponge was collected at a depth of approximately 500 feet. Photo courtesy of Harbor Branch Oceanographic Institution, Inc., ©1998.

Plate XIIb: Untreated human cancer cells stained with fluorescently labeled anti-alpha-tubulin antibody. The individual, green hair-like structures are microtubules which form an organized meshwork or cellular skeleton (cytoskeleton) in cells. Microtubules also assist in the segregation of chromosomes during cell division. Photo courtesy of Harbor Branch Oceanographic Institution, Inc., ©1998.

Plate XIIc: Human cancer cells treated with discodermolide. The microtubule network has become reorganized due to the activity of discodermolide. This results in the formation of microtubule bundles, disruption of cell division, and death of the cancer cells. Photo courtesy of Harbor Branch Oceanographic Institution, Inc., ©1998.

Plate XIII: *Pseudopterogorgia elisabethae*, a Caribbean gorgonian, is the source of potent anti-inflammatory compounds. Photo courtesy of William Fenical, Scripps Institution of Oceanography.

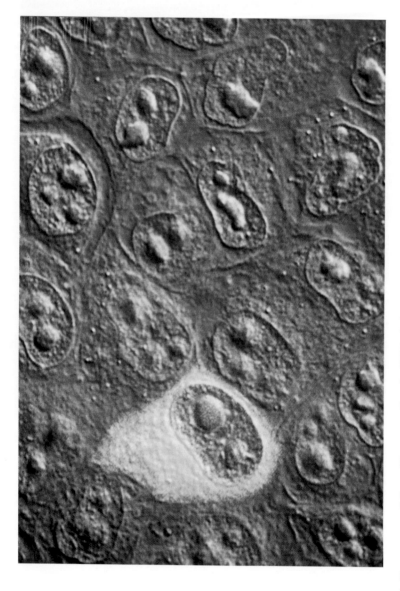

Plate XIV: Two mutants of Green Fluorescent Protein (GFP), derived from a jellyfish, are fused to HIV genes encoding a cytoplasmic protein (green) and a nuclear protein (blue). GFP can be linked to a variety of genes to monitor protein expression and subcellular localization (Stauber et al., 1998).

Plate XV: Photo of the sea urchin *Lytichinus pictus* spawning. The female is inverted on top of the beaker and the plentiful orange eggs drop to the bottom. A spawning male appears to the side of the beaker—the white foam on the top of the urchin contains the sperm. Photo courtesy of Chris Patton, Hopkins Marine Station, Stanford University.

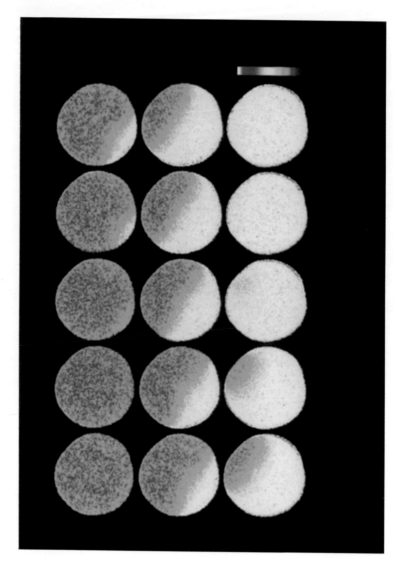

Plate XVI: Confocal microscope images taken at 5-second intervals of a fertilization-induced calcium wave in a *Pisaster ochraceus* starfish oocyte. The color spectrum indicates the relative concentration of calcium where blue-green represents low calcium and yellow-red represents high calcium. Photo provided courtesy of Stephen A. Stricker, Department of Biology, University of New Mexico.

ever, peaked during December, and corresponds with the seasonal name for this phenomenon (El Niño refers to the birth of Jesus). These data could be collected and coordinated with water sampling for zooplankton, *V. cholerae*, nutrients, and chlorophyll, and with epidemiology for cholera incidence.

Storms fueled by the 1997-1998 El Niño increased urban runoff, saturated septic tank leach fields thereby causing overflows, and caused a weakened sewage main to break and spill 120 million gallons of raw sewage into Santa Monica Bay. As a result, public health officials in California closed beaches on 50 days during the period May 1997 through April 1998; this was more closure days than in the three previous years combined for Santa Monica Bay (Alamillo and Gold, 1998).

VECTOR-BORNE DISEASES

Introduction

The ocean influences climate and weather (see Chapter 1) and hence indirectly affect infectious disease agents that are sensitive to changes in temperature and/or rainfall (Pearce et al., 1995). Global warming may affect the incidence of certain protozoal, bacterial, and viral diseases, because increased temperatures tend to change the geographic distribution and reproductive success of many vectors that carry these diseases. For example, higher temperatures may increase the range of some species, but it may also shorten insect survival hence decreasing the transmission of some vector-borne diseases. Indirect effects of climate change may have an impact on human health, but accurate assessment of the risk requires a better understanding of how vectors and their pathogens respond to changes in weather patterns. Indications of how global climate change can affect the spread of vector-borne infectious diseases may be discerned from studying the effects of periodic weather events such as El Niño/ Southern Oscillation (ENSO). Weather events such as ENSO bring extremes of both rainfall and drought and change the capability of disease vectors to spread their pathogens to humans. Some of these changes appear to be due to an increase or decrease in habitat, but the spread of vector-borne diseases is also modified by the effects of these disasters on the water supply and public hygiene. Specific effects of ENSO on several of the vector-borne diseases are described below.

Major Disease Agents and Their Vectors

Malaria: In the aftermath of a severe marine weather event such as a hurricane or monsoon, there is frequently a higher risk of malaria from an increase in standing water breeding habitat and in some cases from a disruption of malaria control measures. Malaria remains one of the most prevalent parasitic diseases in the tropics. The disease agents are the protozoa, belonging to the genus *Plasmo-*

dium, a parasite transmitted by the anopheline mosquito. Worldwide prevalence of the disease is estimated to be 300-500 million clinical cases per year, with 90% occurring in sub-Saharan Africa and most of the remainder in India, Brazil, Sri Lanka, Viet Nam, Colombia, and the Solomon Islands (WHO, 1996). Paradoxically, an increase in malaria transmission may be associated with either drought or rainfall as seen in the Indian subcontinent before the introduction of residual insecticides. In the arid Punjab, rainfall improved the breeding and survival of mosquitoes. However, in the high rainfall area of Sri Lanka, lack of rainfall from the failure of the monsoons led to reduced flow of rivers and the formation of standing pools of water that provided favorable breeding conditions (Bouma and van der Kaay, 1996). Drought associated with the failure of the south-west monsoons are twice as frequent in the year following an El Niño (Dilley and Heyman, 1995) and correlate with past malaria epidemics. Improved predictions of malaria transmission may be possible through the application of hydrological models to estimate available surface water for mosquito breeding (Patz et al., 1998a).

In Venezuela and Colombia the incidence of malaria morbidity and mortality increased substantially (35-37%) in the years following an El Niño. In these countries, the increase correlated more strongly with drought in the first year of the El Niño as opposed to rainfall in the second year (Bouma and Dye, 1997 and Bouma et al., 1997a). It is not clear how drought promotes disease in these countries; however, the authors suggest that in the dry year there is a loss of mosquito predators which allows the mosquito population to increase unchecked in the following wet year.

There has been some question as to whether El Niño conditions gave rise to the epidemics of malaria that occurred during 1983 in Bolivia, Ecuador, and Peru. A review of the data reported by each country on malaria shows that the incidence of this disease began to rise in each of these countries in 1983. However, the overall trend from 1970 to 1996 was an increase in the number of cases reported, while in other El Niño years (1971-1972, 1976-1977, 1991-1992) the incidence of malaria seldom increased over that of previous years. It is known that, during this time, national malaria control programs in Latin America switched from a strategy of rigid eradication to flexible control. This alone could have caused the observed increase. Conversely, a good eradication program may have been masking the impact of El Niño in previous El Niño years. Establishment of an association between an El Niño event and outbreaks of malaria will require future research that emphasizes accurate reporting of outbreaks and consideration of the impact of malaria control measures.

Rift Valley Fever: Rift valley fever (RVF) is a vector-borne viral disease transmitted by the *Aedes* mosquitoes that primarily affects livestock but which also afflicts humans. Incidence of this disease is highly correlated with excessive rainfall, such that high rainfall in an El Niño year may trigger an outbreak of RVF. However, heavy rains are not sufficient to predict an epidemic of RVF. Transmission is also dependent on the presence of both the competent vector

(*Aedes* mosquito) and the pathogen. In 1997, the eastern African countries of Kenya and Somalia experienced 60-100 times the normal rainfall in some areas. This was followed by an outbreak of 89,000 cases of RVF in northeastern Kenya and southern Somalia, possibly the largest ever reported (WHO, 1998). Conversely, Kenya did not experience an outbreak of RVF during the 1982-83 El Niño, despite heavy rainfall.

Dengue: Dengue and Dengue Hemorrhagic Fever (DHF) are caused by four distinct, but closely related, viruses carried by the *Aedes* mosquito. This mosquito has become adapted to urban habitats, breeding in manmade collections of rainwater as well as water stored for drinking and washing. The occurrence of this disease has increased dramatically in the past few decades, from 9 countries prior to 1970 to 41 countries by 1995. Currently, both the *Aedes* mosquito and dengue viruses are endemic to the tropics worldwide. Each year, dependent on the frequency of outbreaks, there are an estimated 100 million cases of dengue fever, and several hundred thousand cases of DHF, a leading cause of death and hospitalization of children in Southeast Asia (Gubler, 1998). As with malaria, it is difficult to demonstrate scientifically that changes in the distribution of dengue are the result of climate variability. PAHO/WHO, in its preliminary study in the Americas, did not find a correlation between national statistics and increased rainfall (PAHO, 1998b). In fact, peaks in reports of dengue did not occur during past El Niños in the Americas.

It is too early in the current El Niño year to evaluate fully the effects on the incidence of dengue; however, 1998 may be the worst year on record for dengue in some areas. In Viet Nam, there were twice as many cases as of June 1998 than in the previous year and 1997 was the worst year since 1991. Similarly, in Brazil there were as many cases of dengue in June as there were for the entire year in 1997, despite the initiation of an eradication program in 1997. Some researchers have proposed models that link potential changes in global climate with changes in the distribution of dengue transmission (Jetten and Focks, 1997; Patz et al., 1998b). However, the overall spread of the disease during the last few decades argues that factors other than weather are responsible for the current increased prevalence of dengue (Gubler, 1998).

There are several other viral diseases that may be affected by changing weather and climate patterns including yellow fever, hantavirus, and viral encephalitis as well as other parasitic diseases such as schistosomiasis and sleeping sickness since the vectors are sensitive to changes in temperature and rainfall (Rogers and Packer, 1993; Stone, 1995).

Evaluating the Impacts of ENSO on Vector-borne Disease

Future research should aim to improve the quality of the indicators and data on disease exposure and health outcome. The poor quality of epidemiologically-based surveillance systems and disease reporting in many countries sometimes

---

BOX 2-1   ENSO Experiment

The ENSO Experiment is a program specifically designed to study the human health impacts of the 1997-1998 El Nino/Southern Oscillation (ENSO) event and determine the potential application of forecasting to public health preparedness. This program is coordinated through the NOAA Office of Global Programs and the International Research Institute for Climate Prediction. The ENSO experiment takes an interdisciplinary approach to examining the ENSO-related changes in weather that have direct and indirect effects on the waterborne and vector-borne infectious diseases that afflict humans.

---

precludes environmental analysis and hinders the planning and implementation of analytic epidemiologic studies. Analysis of aggregated data at the national level by PAHO has not shown a significant correlation between ENSO and communicable diseases. However, because of limitations in the available data, it is possible that correlations were present but not detected. Epidemiological studies and meaningful conclusions on the impact of ENSO or communicable diseases will require a long-term temporal and spatial (geographic) perspective with a sustained quality of data that quick short term investigations will not provide. International cooperation between scientists, countries, and WHO is required to develop the global surveillance system required to respond to these global challenges. Although it is impossible to determine how much human illness is directly caused by ENSO, specific weather patterns give an indication of potential outbreaks of vector-borne diseases. It will be very difficult to separate the effects of El Niño from other factors that impact the transmission of these diseases. One program that is currently being conducted to address these issues is the ENSO Experiment (Box 2-1).

CONCLUSIONS

Given current world population dynamics, including growing populations near coastal waters and increased recreational use of ocean and coastal waters, particularly in developing countries, there are several pressing scientific needs for better understanding the ocean and infectious diseases. The spectrum of infectious diseases in the United States related to several routes of exposure needs to be more thoroughly and comprehensively studied and monitored, nationwide and internationally. These disease categories include: waterborne infectious diseases, vector-borne diseases, infectious diseases related to consumption of tainted seafood, and the infectious diseases related to swimming, boating, fishing, and water use. New tests are needed to detect waterborne pathogens more quickly and efficiently, and more definitive research in the changing epidemiology of infectious diseases is needed. The primary areas deserving attention are listed below.

1.   Newly developed tests, including gene probes for bacteria, viruses, and other pathogens and monoclonal/polyclonal antibody direct detection methods, are available and should be applied to environmental monitoring and quality assessment. However, they must be field tested for application in different geographic locations to expand the database and, if successful, should be adopted by enforcement agencies as standard methods. To achieve this goal, coordination of funding and enforcement agencies, at all levels of government, with industry and academia, will be necessary. Specific suggestions include:

• Evaluate cost/benefits of the use of coliforms as indicators of fecal pollution in estuaries and seawater, as introduction of newer molecular genetic and immunological methods for direct detection and evaluation of pathogens proves applicable.
• Where practical, begin application of such tests for shellfish quality assurance with emphasis on accuracy, precision, speed, cost, and safety.
• Coordinate and integrate environmental monitoring for pathogens with targeted epidemiological investigations to quantify population risk in high-risk areas.
• Evaluate the efficacy and comprehensiveness of current disease surveillance programs aimed at monitoring infectious diseases that are waterborne and related to consumption of seafood. Assess the adequacy of federal programs and the need for coordination with state efforts, especially for coastal states.

2.   Interdisciplinary research should be emphasized/encouraged to assess the potential for global climate change to affect the spread of human diseases. Investigative teams should include, at a minimum, microbiologists, meteorologists, oceanographers, remote sensing experts, epidemiologists, parasitologists, biostatisticians, entomologists, ecologists, botanists, and behavioral scientists. Specific suggestions include:

• Investigate the cause and effect relationships between sea surface temperature, nutrients, plankton, *Vibrio cholerae*, disease incidence, and global climate change.
• Plan and monitor long-term campaigns for the assessment and control of vector-borne diseases that have the potential for being influenced by global climate change.
• Coordinate efforts between the World Health Organization (WHO), U.S. Agency for International Development (USAID), Centers for Disease Control (CDC), U.S. Department of Defense (DOD), and other international organiza-

tions to monitor temporal and geographic disease trends in all nations of the world, so that early warning of emerging diseases can be effective.

• Develop better testing methods for autochthonous pathogens and elucidate their niche (i.e., distribution and role in their natural habitat), so that disease transmission to humans can be prevented.

• Seek advice and participation from key organizations in developing epidemiological methods for application to global climate change issues (EIS Program/CDC; American Teachers of Preventative Medicine; American College of Epidemiology; Society for Epidemiology Research; International Society for Environmental Epidemiology; American Public Health Association; American Society for Microbiology).

3

Harmful Algal Blooms

Blooms of the single cell algae known as phytoplankton are sometimes called red tides, which have been recognized since biblical times. The phytoplankton may become so numerous that they cause the water to become discolored (i.e., red, reddish brown, green, yellow green) and these events are associated with shellfish toxicity and fish kills. Over the past twenty years, phytoplankton researchers (Anderson, 1989; Hallegraeff, 1993; Smayda, 1990; Steidinger and Baden, 1984) have noted an increasing frequency of harmful phytoplankton blooms worldwide. Of the roughly 5,000 phytoplankton species, fewer than 80 are known to be toxic, but once established, some toxic or nuisance blooms may persist because their toxins may inhibit the growth of other phytoplankton or reduce grazing pressure by zooplankton (Turner and Tester, 1989; 1997; Turner et al., 1998). The blue-green algae, or cyanobacteria, are represented by only a few marine genera, but these organisms may pose significant threats worldwide to human health in freshwater systems. The ramifications of harmful/toxic phytoplankton blooms are extensive. The loss of human life and risk of adverse health outcomes are of primary concern. Physicians and public health officials are not always trained to recognize the symptoms of poisoning from exposure to algal toxins. Regional economies are impacted when shellfish resources are tainted and cannot be harvested; mass mortality of finfish and loss of environmental quality result in further economic losses. Marine mammal deaths are linked to the concentration of several phycotoxins within marine food chains, (Bossart et al., 1998; Geraci et al., 1989) and the impact of toxic phytoplankton on non-commercial species can only be conjectured.

The toxic materials produced by harmful algae are "environmental chemicals," toxins that interfere with human and animal metabolism, nerve conduction, and central nervous system processing of information. Our understanding of the mechanisms of algal intoxication depends upon the use of model animal systems, while public health monitoring of seafood and seawater provides information relevant to the health of marine species. For humans, harmful algal blooms cause illness through several routes of exposure. Toxins produced by the algae may contaminate seafood and water, or in some cases become airborne in sea spray. Of the natural marine environmental contaminants that are health risks, harmful algal blooms (HABs) are most prominent.

HARMFUL ALGAL BLOOM HAZARDS IN FOOD

Filter-feeding bivalve molluscs accumulate and concentrate phycotoxins that can be further bioconcentrated as they move through the food chain to top carnivores (Shumway 1990). Human intoxication follows ingestion of tainted shellfish or, in the case of ciguatera, finfish. The severity of symptoms is dependent upon the amount of toxin ingested, the weight and general health of the individuals, and their susceptibility to the toxin. General clinical symptoms of fish and shellfish poisoning include nausea, vomiting, abdominal pain, and diarrhea. Phycotoxins have a high affinity for specific receptor sites leading to critical changes in intracellular ion concentrations of sodium, calcium, or potassium. Consequently, action potential and nerve transmission impulses are affected. HABs are responsible for six different types of seafood poisoning, several of which can be lethal (Table 3-1). Five of these types of seafood poisoning are found in North America on a recurring basis. From 1978 to 1987, more than half of the cases of illness from naturally occurring seafood toxins were the result of harmful algal blooms toxins (IOM, 1991). The first step in determining the public health hazard from an algal bloom is identification of the species and the toxin. This has been especially problematic with *Pfiesteria,* where the alga is identified by electron microscopy, exposure is hard to measure because the toxin appears to be inhaled as an aerosol, and the toxin has not yet been purified and characterized, apparently because the compound is chemically unstable.

Paralytic shellfish poisoning (PSP) occurs from Alaska to Mexico and from Prince Edward Island to Massachusetts. Most often, the toxins are accumulated in bivalve shellfish, but instances of accumulation in mackerel and in carnivorous gastropods have been demonstrated. Neurotoxic shellfish poisoning (NSP) is a hazard in all coastal regions of the Gulf of Mexico and at times on the Atlantic coast as far north as the Carolinas. The NSP toxins accumulate predominantly in shellfish, but recent instances of human intoxication due to consumption of toxic mullet implicates finfish in accumulation (Baden, 1998). Amnesic shellfish poisoning (ASP) causes human illness from Washington State to southern California and on Prince Edward Island. On several occasions, ASP has been documented to accumulate in anchovies; therefore, fish transvection routes to humans must be considered. Diarrheic shellfish poisoning (DSP) has been documented in

TABLE 3-1 Intoxication Syndromes Caused by Marine Toxins Consumed in Seafood

Disease	PSP	NSP	ASP	DSP	Ciguatera	Puffer Fish
Causative Organism	Red Tide Dinoflagellate	Red Tide Dinoflagellate	Red Tide Diatom	Red Tide Dinoflagellate	Epibenthic Dinoflagellate	Bacteria?
Major Transvector	Shellfish	Shellfish	Shellfish	Shellfish	Fish	Fish
Geographic Distribution	Temperate to Tropical Worldwide	Gulf of Mexico, Japan, New Zealand	Canada, NW U.S.	Temperate Worldwide	Sub-Tropical to Tropical Worldwide	Japan, Worldwide
Major Toxin (Number)	Saxitoxin (18+)	Brevetoxin (10+)	Domoic Acid (3)	Okadaic Acid (4)	Ciguatoxin (8+) Scaritoxin, Maitotoxin	Tetrodotoxin (3+)
Neuro-Mechanism	Na+ Channel Blocker	Na+ Channel Activator	Glutamate Receptor Agonist	Phosphorylase Phosphatase Inhibitor	NA+, Ca2+, Channel Activators	Na+ Channel Blocker
Incubation Time	5-30 min	30 min-3 hr	hours	hours	hours	5-30 min
Duration	days	days	years	days	years	days
Acute Symptoms	n,v,d **p,r**	n,v,d, **b, t,** p	n,v,d,**a**, p,r amnesia	**d**, n,v	n,v,d, **t**, p paraesthesias	n,v,**d,p,r,°bp**
Chronic Symptoms	none	none	amnesia	none	paraesthesias	none
Fatality Rate	1-14%	0%	3%	0%	< 1% (0.1-12%)	60%
Diagnosis	clinical, mouse bioassay of food, HPLC	clinical, mouse bioassay of food, ELISA	clinical, mouse bioassay of food, HPLC	clinical, mouse bioassay, HPLC, ELISA	clinical, mouse bioassay, immunoassay	clinical, mouse bioassay, Fluorescence
Therapy	Supportive (respiratory)	Supportive	Supportive (respiratory)	Supportive	Mannitol TCA? Supportive	Supportive (respiratory)
Prevention	red tide and seafood surveillance, report cases	red tide, then seafood surveillance, report cases	seafood surveillance, report cases	seafood surveillance, some red tide, report cases	seafood surveillance, report cases (clusters)	regulated food preparation, report cases

PSP = Paralytic Shellfish Poisoning, NSP = Neurotoxic Shellfish Poisoning, ASP = Amnesic Shellfish Poisoning, DSP = Diarrheic Shellfish Poisoning, Ciguatera (CFP) = Ciguatera Fish Poisoning, Puffer Fish poisoning = Fugu; n = nausea, v = vomiting, d = diarrhea, p = paraesthesias, r = respiratory depression, b = bronchoconstriction, t = reversal of temperature sensation, a = amnesia, °bp = decreased blood pressure. Symptoms in bold indicate pathognomonic symptoms, numbers in () indicate # of natural derivatives (from Baden et al., 1995).

Nova Scotia but is not a current public health problem in the United States. It must, however, be considered as a potential emerging threat in the future. Ciguatera fish poisoning (CFP) is the most common type of seafood poisoning and occurs in virtually all tropical reef regions. International transport of seafood continues to threaten public health and recent implementation of the Hazard Analysis Critical Control Point (HACCP) program by the U.S. Food and Drug Administration (FDA) seeks to address the emerging threat. This program, which was first implemented by General Foods to safeguard foods on space missions, has as its hallmark an identification of all points in a food process whereby its safety or wholesomeness can be compromised. Identification is followed by implementation of strategies to protect or preserve the integrity of the food source.

Paralytic Shellfish Poisoning (PSP)

Paralytic shellfish poisoning is caused by the ingestion of saxitoxin or its derivatives. Saxitoxin was first characterized in 1957 (Schantz et al., 1957) and now includes 21 recognized forms. Each of the known derivatives binds specifically (although with variable affinity) to the voltage-gated sodium channel. These toxins are water-soluble and act primarily on the peripheral nervous system and secondarily on the central nervous system. The onset of symptoms is rapid: gastrointestinal distress, tingling, numbness, and ataxia are typical. Some of the clinically diagnosed individuals die of respiratory failure. As long as medical records have been maintained, human poisoning from eating bivalves has been reported (Shantz, 1984). PSP was recognized by Native Americans before the arrival of European explorers. Several members of Capt. George Vancouver's crew succumbed to PSP while they explored the Pacific Northwest in 1798. Although the toxin is initially accumulated by shellfish, marine mammal deaths have resulted from food chain concentration in mackerel following an unusual temporal passage from red tide to thread herring (Geraci et al., 1989).

Examples of cells containing saxitoxins are *Alexandrium catenella, A. tamaense, Gymnodinium catenatum,* and *Pyrodinium bahamense* var. *compressum.* These species represent sub-Arctic to tropical forms, and most produce cysts or resting stages triggered by temperature or other environmental changes (Anderson et al., 1983). This adaptive strategy also promotes the expansion of PSP blooms from one geographic region to another. Cysts are remarkably resilient and survive transport in ships' ballast water, in the digestive tracts of spat oysters shipped from one region to another, and are sediment-stable for years. Changes in ocean circulation patterns, disturbance of resting cyst populations, and dredging operations can move seed beds of resting cysts to new regions which may be conducive to growth. Until 1970, PSP was known only in the temperate waters of North America, Europe, and Japan; by 1990, PSP was documented in South Africa, South America, the Philippines, Australia, and India (Hallegraeff,

1993). At present, one dinoflagellate species responsible for PSP, *P. bahamense*, is confined to tropical coastal waters of the Atlantic and Indo-Pacific, however, a survey of its fossil cysts indicates a much wider geographic range in the past (Hallegraeff, 1993).

Neurotoxic Shellfish Poisoning (NSP)

Neurotoxic shellfish poisoning is caused by the ingestion of brevetoxin and its twelve different toxic forms. All brevetoxins bind to the voltage-gated sodium channel (Poli et al., 1986) and have the opposite effect of saxitoxin. Instead of acting like a plug on the channel, they act like a door stop and hold sodium channels in their open configuration (Jeglitsch et al., 1998). Uncontrolled nerve impulses result, ultimately leading to respiratory inhibition. Whereas saxitoxin blocks sodium transport, brevetoxin allows unregulated sodium transport.

NSP produces gastrointestinal and neurological symptoms, less severe, but similar to those of ciguatera fish poisoning (see below). Blooms of *Gymnodinium breve*, the dinoflagellate responsible for NSP, are usually marked by large patches of discolored water and massive fish kills. In addition, this unarmored dinoflagellate can be ruptured easily by wave action, whereupon its toxins become aerosolized and cause respiratory asthma-like symptoms. *G. breve* red tides were documented as early as 1844 and their correlation with shellfish toxicity was recognized by 1880. However the identification and chemical characterization of the first of 10 brevetoxins was not completed until 1981, when toxin purification techniques became available (Lin et al., 1981). Toxin structures quickly followed for several other natural brevetoxins.

Historically, the distribution of *G. breve* blooms has been in the Gulf of Mexico, with isolated occurrences recorded along Florida's east coast. However, during the fall and winter of 1987-88 there was a large, persistent *G. breve* bloom in the coastal waters of North Carolina, a range extension of 800-900 km for this species (Tester et al., 1989). Forty-eight cases of NSP were documented and more than $24 million dollars was lost to the local economy when many shellfish harvesting areas were closed for the entire season (Tester and Fowler, 1990). Subsequently, an explanation for this unusual event was uncovered when this dinoflagellate was found in low but consistent numbers in the Gulf Stream (Tester et al., 1991). The shoreward intrusion of warm water from meanders of the Gulf Stream, seen in Plate VIII, transported *G. breve* to the nearshore waters of North Carolina.

In 1996, following an extensive Florida red tide, over 150 West India manatees died as a result of toxin exposure (Bossart et al., 1998). Other toxic species related to *G. breve* are known to cause fish kills, shore bird deaths, and shellfish toxicity in Japan, New Zealand, and possibly South Africa. The Japanese and New Zealand species produce toxins similar to brevetoxin.

Ciguatera Fish Poisoning (CFP)

Ciguatera fish poisoning is an operational term that includes all lipid-soluble toxins that accumulate in tropical reef fish flesh, which when consumed lead to a debilitating disease characterized by reversal of temperature sensation, chronic pain and numbness in the extremities, joint and bone pain. In severe cases, these symptoms are known to persist for weeks to months and, in a few isolated cases, neurological symptoms have persisted for several years. In other cases, patients have experienced a recurrence of neurological symptoms months to years after recovery (U.S. FDA, 1997). Several ciguatoxins have been isolated and a variety of other toxins are thought to contribute to the syndrome. Ciguatoxin, isolated from Pacific moray eel tissue, binds to the same site of voltage gated sodium channels as does brevetoxin.

CFP was first recognized in the 1550s in the Caribbean (Martyr and Novo, 1912), but the causative agent was not identified until the mid-1980s (Carmichael et al., 1986; ILO, 1984; Sakamoto et al., 1987). CFP has a pantropical distribution between 34 S and 35 N and is known from the Caribbean basin, Florida, the Hawaiian Islands, French Polynesia, and Australia (Anderson and Lobel, 1987). It has been associated with a suite of at least 6 toxins produced by a multispecies assemblage of benthic, (sessile, epiphytic) dinoflagellates, including *Gambierdiscus toxicus*, some *Prorocentrums, Ostreopsis,* and *Coolia.* Ciguatoxin structures resemble brevetoxin, and their molecular mechanism of action is identical. These toxins are bioconcentrated by higher carnivores, especially reef fish, which may remain toxic for more than 2 years after becoming contaminated (Helfrich et al., 1968). There is mounting evidence that Pacific and Caribbean ciguatera toxins are different chemical entities and many investigators believe there are some elusive toxins within the "ciguatera" operational definition that have not yet been isolated.

Worldwide, 50,000 victims are stricken annually (Bomber and Aikman, 1988/89) with CFP; cases per thousand residents vary between 3-9 in the Caribbean to 5-13 in French Polynesia. It is estimated that only 20-40% of the cases are reported. In the acute phase of CFP, gastrointestinal distress is followed by neurological and cardiovascular symptoms that can be, but rarely are, fatal. A chronic phase can persist for weeks, months, or years (Freudenthal, 1990). There is no antidote to CFP and supportive therapy is the rule. In extreme cases of CFP, death through respiratory paralysis may occur within 2-24 hours of ingestion. Repeated exposure to ciguatoxins exacerbates the symptoms, therefore, CFP is considered a major health and economic problem in many tropical islands where fish is a large part of the diet. CFP is one of the most important constraints to fisheries resources development in these regions (Olsen et al., 1984) and also poses a threat to uninformed tourists (Freudenthal, 1990). CFP accounts for over half of all seafood intoxication (IOM, 1991)

Diarrheic Shellfish Poisoning (DSP)

Diarrheic shellfish poisoning is due to the consumption of okadaic acid and two related derivatives. Unlike the toxins that interact with nerve channel proteins, this toxin group inhibits protein phosphatases, a group of enzymes responsible for smooth muscle function, for regulation of cell division in vertebrates, and for overall phosphate metabolism. Diarrhea and tumor promotion are two toxic effects ascribed to okadaic acid.

DSP was first reported from Japan in 1976, (Yasumoto et al., 1980) where okadaic acid produced by several species of the dinoflagellate *Dinophysis* and *Prorocentrum* was found to be the cause. DSP is not fatal, recovery is within three days with or without medical treatment and its symptoms are easily mistaken for bacterial gastric infections. Over a 5-year period (1976-1982), 1,300 DSP cases were reported in Japan; in 1981 5,000+ cases were reported in Spain; in 1983 3,300+ cases were reported in France (Hallegraeff, 1993). DSP has been documented in Japan, Europe, Chile, Thailand, and New Zealand, but prior to 1990 DSP was not known to occur in North or South America. Then, in 1990 and 1992, DSP occurred along the southern coast of Nova Scotia (Quilliam et al., 1993). DSP was also documented in Uruguay in 1992 (Mendez, 1992). Some consider DSP to be the most widespread phytotoxin-caused seafood illness. This is particularly significant because of recent findings indicating that okadaic acid is mutagenic (Anune and Undestad, 1993). Although DSP-producing species of phytoplankton occur throughout all temperate coastal waters of the United States, no outbreaks of DSP have been documented in U.S. waters.

Amnesic Shellfish Poisoning (ASP)

Amnesic shellfish poisoning is due to the accumulation of domoic acid by shellfish. Domoic acid binds to a specific subset of glutamic acid brain receptors known as the kainate receptor. Normally, this receptor, in part, functions in establishing short- and long-term memory. Impaired, intoxicated individuals can die if the dose is sufficient or experience permanently impaired memory function. ASP was recognized for the first time in 1987 on Prince Edward Island when over 100 acute cases and 4 deaths resulted from consumption of blue mussels (Bates et al., 1989). Subsequent studies of this illness revealed that the neurotoxin domoic acid, produced by a diatom, *Pseudo-nitzschia multiseries*, caused the ASP outbreak. Typical symptoms of severe cases include gastroenteritis followed by dizziness, headache, seizures, disorientation, short-term memory loss, and respiratory difficulty.

In the Bay of Fundy, generally two blooms of *Pseudo-nitzschia* occur each year; one when the water temperature warms to about 10 °C and the second occurs later, following the highest water temperatures of the year in late August (Martin et al., 1993). Despite the annual blooms of the diatoms that cause ASP in Canadian waters, contaminated shellfish have been kept off the market by vigilant management practices. Hence, public confidence in the local mussel industry is high; mussel harvests there now exceed the 1987 levels (Wood and Shapiro, 1993).

In the fall of 1991, domoic acid was detected in dead sea birds in Monterey Bay, California. They had been feeding on anchovies that had ingested *Pseudonitzschia australis*, another source of domoic acid. Further tests found domoic acid present in razor clams and crabs from Oregon and Washington; subsequently, both recreational and commercial fisheries were closed (Wood and Shapiro, 1993). As recently as May 21-31, 1998, the death of 50 California sea lions in the Monterey area was caused by domoic acid. Although no known human intoxication resulted from either of these incidents, it was a clear warning that domoic acid can accumulate in marine food chains.

Cyanotoxins

Cyanotoxins (i.e., alkaloid neurotoxins, hepatotoxins) are produced by some species (or strains) of all the common freshwater genera of blue-green algae, also known as cyanobacteria (e.g., *Anabena, Aphaanizomenon, Microcystis, Nodularia, Nostoc, Oscillatoria*) (Carmichael, 1992) and several species of marine cyanobacteria, including *Trichodesmium thiebautii* (Guo and Tester, 1994; Hawser et al., 1991). These cyanotoxins produce intermittent but repeated cases of animal poisonings in many areas of the world. Poisoning cases, known since the late 19th century, involve sickness and death of livestock, pets, and wildlife following ingestion of water containing toxic algae or the toxin(s) released by the aging cells (Charmichael, 1992). No acute lethal poisoning of humans by consuming foods containing freshwater cyanobacteria, such as occurs with paralytic shellfish poisoning, has been confirmed. There are no known food vectors, such as shellfish, to concentrate toxins of freshwater cyanobacteria in the human food chain. However, the decreasing water quality and increasing eutrophication of freshwater supplies mean that large growths or waterblooms of cyanobacteria are becoming more common (Paerl, 1988), increasing the probability that humans could be exposed to a toxic dose of these algae (Charmichael, 1992). Inadvertent poisoning by the freshwater blue-green algal toxin microcystin (a functional homolog of okadaic acid) in kidney dialysis machines in Mexico has been confirmed as the cause of 30 deaths. Incomplete municipal water treatment was identified as the culprit.

Other Human Routes of Exposure

In the preceding section, the concern has been on the accumulation of natural toxins in seafood. This concern is well-founded, for the adulterated seafood can have worldwide distribution as a result of international food marketing practices. But, in localized areas where the blooms occur, a variety of physical perturbations can result in the ejection of active toxin into the air as aerosols. The agents are neither noxious gases nor are they vapors. Rather, the toxins become associated with micro-particles of water resulting from the bursting of bubbles at the water surface. Depending upon the length of time in the airborne state, water can evaporate from the aerosol particles leaving dry salt with toxin coating the particle or entrapped within it. Only two types of toxic blooms are known to affect people by inhalation: *Gynmodinium breve* (Florida red tide) and *Pfiesteria*.

The classic example of noxious natural toxins being liberated in aerosolized form is the Florida red tide and its brevetoxins. This phenomenon has been described for over 100 years. During Florida red tide, persons on the beaches experience a tightness of breath, mucous discharge from the nose, coughing and sneezing, and tearing eyes and a burning sensation in mucous membranes. Airborne toxin can travel far inland, and removal of persons to a toxin-free environment or the donning of a particle filtration face mask relieves the debilitating symptoms. It is the same toxins, or brevetoxins, that cause this effect as cause the NSP described earlier. These toxins cause their noxious effects at concentrations in air in the femto- to pico-gram per liter range.

Pfiesteria piscicida (Burkholder et al., 1993), *Cryptoperidinopsis brodyii* nov. gen. nov. sp. (Steidinger, Landsberg, and Truby, In Review) and several other *Pfiesteria*-like heterotrophic dinoflagellates have been linked to lesioned fish kills in eastern United States coastal waters (Burkholder et al., 1993). Characteristically these events occur in brackish water (<15 ppt salinity) during the warmest part of the year in slow moving waters with lower oxygen content. While reports suggest that exposure to estuarine waters in Maryland during fish kills in the late summer and fall 1997 caused neurocognitive deficits in several individuals, the Centers for Disease Control and Prevention is using the term "estuary associated syndrome" to describe the phenomenon associated with such exposure. This is based on a review of the Maryland findings and the inability to attribute the adverse effects to a specific dinoflagellate or toxin (Smith and Music, 1998). However, another recent report correlated the level of exposure to waters containing *Pfiesteria* or *Pfiesteria*-like dinoflagellates with the likelihood of developing learning and memory difficulties (Grattan et al., 1998). A number of different genera may be involved and several may produce toxins. Unpublished reports of two *P. piscicida* toxins suggest that one is a water-soluble neurologic agent and another is a lipid-soluble dermonecrotic agent. The present difficulty in the identification and characterization of these toxins might be explained if these particular toxins degrade comparatively rapidly in the environment. There is no evidence to date of food chain contamination from *P. piscicida* or *Pfiesteria*-like heterotrophs, unlike the other heat- and cold-stable dinoflagellate toxins known to cause human illness primarily via consumption of contaminated shellfish or finfish. *Pfiesteria* and *Pfiesteria*-like species are the subject of intensive investigation by researchers from Maryland to Florida of the potential causal links between the presence of these heterotrophic dinoflagellates and fish kills, fish lesions, and human health (Buck et al., 1997).

The recent identification of a fungus in menhaden fish lesions, *Aphanomyces*, by a team at the U.S. Geological Survey (USGS) implicates it in lesioned fish episodes and makes the *Pfiesteria* story much more complicated. The USGS team found the fungus in 95 percent of lesions on fish taken from the Chesapeake Bay in 1997 during an outbreak of *Pfiesteria* (Vicky Blazer, fish pathologist at the USGS, personal communication).

RESEARCH REQUIREMENTS AIMED AT DIAGNOSTICS, THERAPEUTICS, AND PREVENTION

The ecology of each harmful algal bloom organism is different. Many are photoauxotrophs, that is, they are photosynthetic and carry out their lives by fixing carbon and utilizing only small amounts of simple nutrients, including nitrogen and phosphorus, and essential vitamins and minerals. Some are capable of limited heterotrophy, utilizing more complete carbon compounds or consuming other organisms like bacteria. Yet others, like *Pfiesteria, Cyptoperidiniopsis, Amyloodinium,* and the rest of the truly heterotrophic dinoflagellates lead a predatory or parasitic existence. The mechanisms by which each organism engulf or acquires its nutrition requires further study. The factors that lead to initiation of a bloom, maintenance of the bloom, and termination of the bloom are not completely understood for any species.

Plate IX shows the occurrence of HAB-related events in the United States before and after 1972. This figure illustrates the increase in the range of HABs, but the frequency of events also appears to be higher. The reasons for this expansion are unknown, but possible explanations include natural mechanisms of species dispersal and human-related phenomena such as nutrient enrichment, climatic shifts, and more accurate reporting of HAB events. As shown on the map in Plate IX, virtually all coastal regions of the United States are now subject to a variety of HAB events. Closer monitoring of the location and frequency of blooms, as well as the physical, biological, and chemical characteristics of the affected bodies of water (as called for in ECOHAB, Box 3-1) will help to resolve why HAB events are increasing in frequency and range. In addition, this monitoring will allow earlier notification of public health authorities so that they can act to reduce exposure of the public to algal toxins.

The detailed mechanism of toxicity is known for only one of the HABs, saxitoxin, and in fact there is still active debate about the microscopic site of interaction of saxitoxin with nerve membranes. It is known that saxitoxin and tetrodotoxin are nearly identical in their effects, that brevetoxins and ciguatoxin are very similar, and that the microcystins and okadaic acid behave in a similar

BOX 3-1 ECOHAB

ECOHAB, the Ecology and Oceanography of Harmful Algal Blooms is a national research agenda to understand, predict and mitigate the causes and consequences of blooms of harmful or toxic algae. The ECOHAB program is a partnership among NOAA, the National Science Foundation (NSF), the Environmental Protection Agency, the Office of Naval Research, the National Aeronautics and Space Administration, and the Department of Agriculture. The objective of this program is to investigate the physical, chemical, and biological oceanographic properties important for understanding the populations dynamics of harmful algal species and the environmental consequences of harmful algal blooms. The results of these studies will form the basis for reducing the impacts of harmful algae on public health, marine ecosystems, and coastal economies. This program began as the result of a workshop co-sponsored by NSF and NOAA where a scientific consensus for the steps necessary to address the HAB problem was developed and then drafted as the National Plan for Marine Biotoxins and Harmful Algae.

toxicologic fashion. This information is essential for development of any therapeutic strategies. Thus, although HABs produce chemicals of high toxicity, more information is needed on exactly how they work at the cellular and molecular level. Success in this area will lead to improved diagnostics, development of potential therapies, and early warning systems for prevention and monitoring purposes.

CONCLUSIONS

Algal toxins in food, water, and the air affect the health of humans and animals. Also, harmful algal blooms disrupt the economies of coastal communities through the closure of fisheries affected by algal toxins. The increasing reports of bloom occurrences and intensities worldwide has brought this issue into prominence; there is concern that HABs signal an underlying deterioration of the marine environment. However, the conditions that provoke algal blooms are not well understood and appear to vary among different species of algae. Several strategies to address these concerns are as follows:

- determine the physical, chemical, and biological factors that promote blooms of specific harmful algal species through increased monitoring of environmental conditions,
- improve methods for accurately identifying the algal species responsible for a bloom,

- determine the molecular mechanisms for the action of natural marine toxins on animals to help develop antidotes, to improve detection, and to understand the pharmacology of toxins,
- ensure safety of seafood through development of cost-effective methods for detecting algal toxins in seafood,
- document the incidence of toxin-related illness in coastal areas and among travelers to high risk areas. There is a need for comprehensive assessment and reporting of the temporal and geographic distribution of algal blooms and associated human illnesses,
- train public health authorities in coastal states to recognize and respond to outbreaks of toxin-related illness.

Part II

———————◯———————

Value of Marine Biodiversity to Biomedicine

In Part I, threats to human health from the ocean were defined and identified as areas that would benefit from greater interaction between the oceanographic and medical communities. For example, research on the ecology of harmful algal blooms addresses problems relevent to both public health and environmental health.

Sentinel species like marine mammals, fish, and birds provide initial indicators of environmental problems such as a harmful algal bloom. Tests to detect Florida red tide and ciguatera, developed for emergency room use, were used to identify red tide brevetoxin as the reason for manatee deaths in 1996. The cause of gannet and other sea bird deaths in 1995 was linked to paralytic shellfish poisoning, using techniques developed to study human nerve function. Using tests developed for molecular brain research, the lethal agent responsible for mortality of sea lions in Monterey Bay was identified as the algal toxin that causes amnesic shellfish poisoning. Animal model system work, funded by the National Institutes of Health (NIH), has provided answers to how toxins are accumulated and how these toxins affect nerves and metabolism. This knowledge provides clues to potential antidotes, therapies, and treatments. Hence, studies of marine toxins not only help prevent human illness and identify environmental problems but also provide insight into human biology and suggest new approaches for treating human diseases.

In the second half of this report, research areas are defined that establish the basis for a unique partnership—a cross-fertilization that provides for an interactive approach to the ocean and human health. Ways in which issues common to different areas of marine research areas will help provide answers or approaches of value to the others are summarized in the table below:

71

TABLE PART II   Interactive Elements of HABs, Marine Models, and Drug
Discovery

	Harmful Algal Blooms	Drug Discovery Programs	Marine Models of Human Disease
Environmental chemicals that affect man	Known and emerging toxins	Surveys for new agents	Test system for identification, including sentinels
High throughput assays	Needed	Available	Receptor basis for development
Mechanism of action	Known but need detail	Bioassay-guided	System for deciphering
Model system	Need sensitive and specific models	Models needed for specific actions	Disease-based
Therapeutics	Need for each toxin type	Potential for specific diseases	*In vivo* system for development

The research described in the following chapters falls under the category of marine biotechnology. Several previous reports have argued that the investment in marine biotechnology in the United States should be higher (NRC, 1994a; Zilinskas. 1995). For example, approximately one patent is issued for every $1.1 million spent on marine biotechnology, an indication of the high productivity of this research. Despite this success, it is estimated that only 1.1% of the nation's biotechnology budget was spent on research in marine biotechnology in 1992 (Zilinskas, 1995). Also, the absolute level of funds spent for marine biotechnology has been low compared to the investment made in other countries; in 1991, the U.S. spent between 7-10% of the amount spent in Japan on marine biotechnology (Zilinskas, 1995). The following chapters illustrate the potential of two areas of marine biotechnology, marine natural products and marine biomedical models. In Chapter 4, the principle of using marine biotoxins and other compounds from marine organisms as molecular probes in biomedical research is discussed in the larger context of how marine natural products provide a new source of molecular diversity that will be of value in developing new pharmaceuticals. Chapter 5 illustrates how basic research on marine organisms offers insights into disease processes that occur in humans.

4

Marine-Derived Pharmaceuticals and Related Bioactive Agents

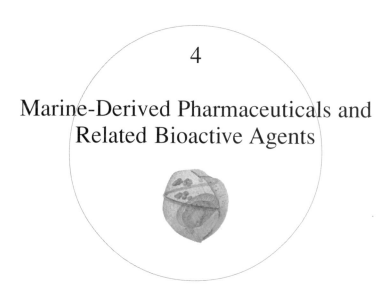

One of the major accomplishments of the 20th century is the development of modern pharmaceuticals. Since their emergence early in the 20th century, drugs such as penicillin, streptomycin, and vincristine, among others, have contributed significantly to the management of human disease. New drug therapies have extended human life span and improved the quality of life. In this regard, society has become more and more reliant upon the availability of safe and efficacious pharmaceutical products.

Nature has been the traditional source of new pharmaceuticals. Today, over 50% of the marketed drugs are either extracted from natural sources or produced by synthesis using natural products as templates or starting materials. Since ancient times, early societies used natural medicines, generally as crude extracts from plants, to treat infection, inflammation, pain, and a variety of other maladies. Even today, in many parts of the world, natural medicines provide the only treatments available. Investigation of these "natural" ethnobotanical preparations led to the isolation of compounds whose beneficial properties have provided the foundation of the current pharmaceutical industry.

Of course, times have changed and science in both industrialized and developing nations has become much more sophisticated in its approach to drug discovery. Complex programs have been initiated to investigate diseases based on their fundamental biochemical and molecular causes. In some cases, it has been feasible to design drugs based on knowledge of an appropriate biochemical target. In other instances, new drugs have been discovered by modern high-throughput screening, a process in which thousands of natural or synthetic chemicals are tested in automated pharmacological bioassays. All of these approaches have a place in the discovery process, but without a natural product to lead science to its biological target, many drugs would never have been developed. It appears that a diverse approach, involving all these methods, is the most effective approach. It is also clear that natural products, which have evolved over millions of years of selective pressures, provide one of the most important components of this process.

There is a continual need for new therapeutic agents, especially to treat a large variety of diseases for which there are no effective therapies. Many forms of cancer, viral and fungal infections, inflammatory diseases, and neurodegenerative diseases cannot be treated successfully. Development of resistance of pathogenic microorganisms to antibiotics and cancer cells to antitumor drugs requires the compensatory generation of new drugs. The pathogens responsible for malaria and tuberculosis have developed resistance to most available drugs. Thus, heavy investment in the development of new antibiotics will be necessary in the next millennium.

This chapter will address the issue of new drug discovery. As in the past, natural products will be an important source of new therapeutics, but it is necessary to identify sources of new biological activities and chemical structural diversity. As described below, the biological diversity of the ocean offers great promise as a source of drugs for the future.

THE MARINE ENVIRONMENT AS A SOURCE
OF CHEMICAL DIVERSITY

Representatives of every phylum are found in the sea; twelve phyla are exclusively marine. The ocean contains more than 200,000 described species of invertebrates and algae (Winston, 1988), however, it is estimated that this number is but a small percentage of the total number of species that have yet to be discovered and described. Conservative estimates suggest that oceanic subsurface bacteria could constitute as much as 10% of the total living biomass carbon in the biosphere (Parkes et al., 1994). From a relatively small number of these species that have been studied to date, thousands of chemical compounds have been isolated (Ireland et al., 1993). Moreover, only a small percentage of these compounds has been tested in clinically relevant bioassays. The ocean represents a virtually untapped resource for discovery of novel chemicals with pharmaceutical potential.

Marine plants, animals, and microbes produce compounds that have potential as pharmaceuticals. These "secondary metabolites," chemicals that are not needed by the organism for basic or primary metabolic processes, are believed to confer some evolutionary advantage. Because many of these plants and animals live in densely populated habitats (Plate X), are non-motile, and have only primitive immune systems, they have evolved chemical compounds to help defend against predators (Paul, 1992), to attract or inhibit other organisms from settling or growing on them (Pawlik, 1993), and to provide chemical cues to synchronize reproduction among organisms that expel their eggs and sperm into the water (Morse, 1991). The mechanisms by which they prevent encroachment or predation interact with the same or similar enzymes and receptors that are involved in human disease processes. For example, many natural products have been identified that inhibit cell division, the process that is the primary target of many anticancer drugs. In most cases, there is a greater understanding of the effect of the natural product on human disease processes than of the function in the marine organism from which it was isolated.

The marine environment became a focus of natural products drug discovery research because of its relatively unexplored biodiversity compared to terrestrial environments. The potential of marine natural products as pharmaceuticals was introduced by the pioneering work of Bergmann in the 1950s (Bergmann and Feeney, 1951; Bergmann and Burke, 1955), which led to the only two marine-derived pharmaceuticals that are clinically available today. The anticancer drug, Ara-C, is used to treat acute myelocytic leukemia and non-Hodgkin's lymphoma. The antiviral drug, Ara-A, is used for the treatment of herpes infections (McConnell et al., 1994). Both are derived from nucleosides isolated from a shallow-water marine sponge collected off the coast of Florida.

Marine sponges are among the most prolific sources of diverse chemical compounds with therapeutic potential (Plate XI). Of the more than 5000 chemical compounds derived from marine organisms, more than 30% have been isolated from sponges (Ireland et al., 1993). Sponges occur in every marine environment, from intertidal to abyssal regions, in all the world's oceans, and they produce a greater diversity of chemical structures than any other group of marine invertebrates. Other marine sources of bioactive molecules with therapeutic potential are bryozoans, ascidians, molluscs, cnidarians, and algae. Several strains of phytoplankton, especially cultured species of diatoms, have been described as exhibiting antibacterial and antifungal activity (Viso et al., 1987). However, the levels of activity are low and hence the active compounds have not yet been isolated or characterized.

THE DISCOVERY AND DEVELOPMENT OF MARINE PHARMACEUTICALS: CURRENT STATUS

Since the mid-1970s, academic, government, industrial, and private research laboratories have devoted varying levels of effort to the discovery of marine-derived pharmaceuticals. The major emphasis has been on the discovery of anticancer compounds, due in large part to the availability of funding to support marine-based drug discovery. The National Cancer Institute (NCI) has led this effort through its aggressive programs to support both single-investigator and multi-institutional cancer drug discovery research. As a result, several marine-derived compounds are in clinical trials for the treatment of cancer.

Bryostatin, isolated from the bryozoan *Bugula neritina*, is a polyketide with both anticancer and immune modulating activity (Kalechman et al., 1982; Pettit et al., 1996; Philip et al., 1993; Suffness et al., 1989). Its mechanism of action is through activation of protein kinase C mediation of cell signal transduction pathways. This compound is currently in Phase II clinical trials for non-Hodgkin's lymphoma, chronic lymphocytic leukemia, and multiple myeloma through a Cooperative Research and Development Agreement (CRADA) between the NCI and Bristol-Myers Squibb.

Ecteinascidin 743, a complex alkaloid derived from the ascidian *Ecteinascidia turbinata* (Rinehart et al., 1990; Wright et al., 1990) and licensed by the University of Illinois to PharmaMar S.A., is in Phase I clinical trials for ovarian cancer and other solid tumors in the United States and Europe.

Discodermolide, a polyketide isolated from deep-water sponges of the genus *Discodermia* (Plate XIIa; Gunasekera et al., 1990) is a potent immunosuppressive and anticancer agent which inhibits the proliferation of cells by interfering with the cell's microtubule network (Plate XIIb and c; Longley et al., 1991a and b; ter Haar et al., 1996). This compound may be effective against breast and other types of cancer that have become resistant to other microtubule disrupting drugs. Discodermolide has been licensed by Harbor Branch Oceanographic Institution (Fort Pierce, FL) to Novartis Pharmaceutical Corporation, and is in advanced preclinical trials.

Another promising sponge metabolite in advanced preclinical trials at the NCI is halichondrin B, derived from a New Zealand deep water sponge, *Lissodendoryx* sp. (Hirata and Uemura, 1986; Litaudon et al., 1994). Like discodermolide, halichondrin B blocks cell division by disruption of microtubule structure.

In each of these cases, however, bulk supply of the chemicals for continued clinical development is a problem. It is often neither economically nor ecologically feasible to rely on large-scale collections of the source organisms from their natural habitats for supply of marine drug candidates. Whether a drug company decides to support the clinical development of a new drug is dependent on identifying an adequate supply of the source material. This could be through culture of the organism or through synthesis of the compound using an economically feasible, industrial-scale process. Research is in progress on options for biological supply (e.g., aquaculture, cell culture, microbial fermentation, and genetic engineering) to address this critical issue in the development of chemicals from natural sources. Unfortunately, neither the NIH nor most drug companies are prepared to invest funds in basic research to develop general models for biological supply of marine natural products with therapeutic potential.

Despite the emphasis on identifying new anticancer compounds, marine natural products have also been found to have other biological activities, including mediation of the inflammatory response. The pseudopterosins are glycosides derived from the Caribbean soft coral *Pseudopterogorgia elisabethae* (Plate XIII; Look et al., 1986; Roussis et al., 1990). These are in advanced preclinical trials as anti-inflammatory and analgesic drugs.

A number of marine-derived compounds have been discovered with antiviral and antifungal activity. Indeed, one of the two clinically-available marine-derived drugs is used for the treatment of herpes infections. Although no marine natural products are currently in clinical trials as treatments of infectious diseases, there is high potential for future development.

MARINE MICROORGANISMS AS A NOVEL RESOURCE FOR NEW DRUGS

Since the discovery of penicillin in the late 1920s, terrestrial, mainly soil-derived microorganisms have provided the single most important resource for discovery of new drugs. Over 120 microbially-produced drugs are in clinical use today to treat infectious diseases, cancer, and to facilitate organ transplantation by suppression of the immune response. Examples of these highly used drugs are the antibiotics, such as the penicillin, cephalosporines, streptomycin, and vancomycin, the cancer drugs actinomycin and mitomycin, and the immunosuppressant drug cyclosporin. Not only have microorganisms been a tremendous source of biodiversity and chemical diversity, but their capacities to produce highly complex molecules from common nutrients in fermentation culture have led to their widespread use in the economic, industrial-scale production of drugs. Today, the pharmaceutical industries worldwide (but particularly in the United States and in Japan) continue to rely on microorganisms as the single most useful source for natural product drugs.

But there are growing concerns about the continuance of this historic approach to drug discovery. Over the past 50 years, soil-derived microorganisms have been extensively investigated from virtually all terrestrial regions of the Earth. Literally millions of isolated microbial strains have been extensively evaluated in a large variety of pharmacological bioassays. Although these microorganisms were at one time a rich source of novel chemical compounds, much of their chemical diversity has been exploited. New antibiotics are not being developed as before. Anticancer drugs, possessing radically new chemical formulations, are not being found at the rate once observed. Depending upon the area of investigation, it has been observed that greater than 95% of the "active" molecules identified are actually molecules discovered in the past. This extensive duplication of discovery has required the development of sophisticated "dereplication" schemes, which are designed to quickly discard well-known substances that frequently recur. More important, the costs of new discoveries have escalated. These costs are so high that the pharmaceutical industry has invested in a more cost efficient process, "high-throughput screening," which allows literally hundreds of thousands of samples to be evaluated in weeks to months, rather than years.

Most of the Earth's microbial diversity is found in the ocean. In addition to the typical organisms found on land, many classes of microorganisms exist only in the sea. The photosynthetic microorganisms alone comprise over 12 plant divisions. With the inclusion of the known unique adaptations of microorganisms to high salt environments and high hydrostatic pressure, the immense diversity of the microorganisms in marine habitats becomes apparent. However, the pharmaceutical industries have not taken advantage of this enormous resource. The reason rests with the tendency of researchers to explore familiar and more readily accessible microorganisms. In the 1950s, it was reported that fewer than 5% of the marine bacteria present in environmental samples could be cultured. This led to the widespread belief that marine microorganisms were simply unculturable. This, together with the traditional lack of cross-disciplinary interaction between marine and medical microbiology, resulted in a great hesitation to embark on an aggressive marine-oriented program. With no guarantees that any new drugs will be found, the risk was difficult to justify. However, times have changed and marine microorganisms can now be cultivated successfully (Davidson, 1995; Fenical, 1993; Kobayashi and Ishibashi, 1993; Okami, 1993). The microbial diversity of the marine environment has thus become available for scientific study, and new programs are emerging worldwide. In the United States and Japan, intense focus on coastal sediments and organisms associated with marine invertebrates has resulted in a growing number of papers documenting the production of novel, bioactive metabolites. Both bacteria and fungi are now the target of biomedical study, and fascinating reports of novel metabolites are becoming more and more common. Bacterial samples from coastal sediments, when grown under saline conditions, have yielded new antibiotics, antitumor, and anti-inflammatory compounds (Pathirana et al., 1992; Trischman et al., 1994a, b). Similarly, when the surfaces of marine plants and the internal tissues of invertebrates have been sampled, bacteria and fungi have been discovered that also produce bioactive compounds with pharmaceutical potential. Marine fungi, in particular, seem to be the focus of increasing interest (Belofsky et al., 1998; Cheng et al., 1994; Kakeya et al., 1995; Numata et al., 1992; Takahashi et al., 1995).

A unique source in the world's oceans, the deep ocean and the geothermal vents, is now becoming the focus of considerable microbiological interest. Microbiological studies of the deep sea environment have shown the presence of obligate barophiles, bacteria which require pressures as high as 600 atmospheres for growth to occur (Yayanos, 1995). These highly adapted marine microorganisms represent a biomedical resource of unknown magnitude, but great promise, as demonstrated recently when unusual compounds produced by bacteria retrieved from deep sea drilling cores were shown to inhibit colon tumor cell growth and prevent replication of HIV, the AIDS virus (Gustafson et al., 1989).

THE MARINE ENVIRONMENT AS A SOURCE
OF MOLECULAR PROBES

One important application of the many bioactive compounds derived from the marine environment is their use as molecular probes, molecules broadly defined as non-drug substances which can be used to probe the foundations of important biochemical events. Molecules such as the potent marine neurotoxins, tetrodotoxin, saxitoxin, conotoxin, lophotoxin, and others, have been instrumental in defining the functions and overall structures of the membrane channels which facilitate nerve transmission. Knowledge of the function of these neurotoxins has allowed drugs to be designed and targeted to those sites of nerve transmission. Other examples are the discovery that the dinoflagellate toxin, okadaic acid, shows potent and selective inhibition of phosphatases; the utilization of the anti-inflammatory agent, manoalide, as a selective inhibitor of the inflammation enzyme phospholipase A_2 (Glaser and Jacobs, 1986); the use of latrunculin A, jaspamide (jasplakinolide), and swinholide A as selective binding agents to the intracellular actin network (Bubb et al., 1995; Matthews et al., 1997; Senderowicz et al., 1995); and the recent discovery of the unique sponge metabolite, adociasulfate-2, which selectively inhibits the intracellular molecular motor protein kinesin (HHMI, 1998). The importance of molecular probes in resolving the complexities of diseases and cellular processes has often outweighed any value that they would have as commercial drugs.

Marine natural products have not only contributed probes for studying specific cellular proteins and enzymes, but they have also provided visual markers for proteins specified by antibodies, for cellular events mediated by calcium, and for elucidating mechanisms of tissue-specific gene expression. Antibodies are an indispensable tool of molecular biology and biomedicine because they can be used to identify specific biomolecules. However, they must be coupled to a reporter molecule, usually an enzyme with a colorimetric substrate or a fluorescent compound. Phycoerythrin, a fluorescent protein isolated from red algae, is crosslinked to antibodies for use as an indicator in many immunological assays. In the algae, phycoerythrin functions in light harvesting during photosynthesis. For this role, the protein shows optimization of both absorption and fluorescence resulting in high quantum yield while showing minimal dependence on pH or ionic conditions. (Glazer, 1988; 1989) These properties make it ideal for assays requiring high sensitivity (quantum yield is 30-100 times greater than the chemical dyes fluoresein and rhodamine); therefore phycoerythrin-conjugated antibodies are a favored reagent for use in flow cytometry, a common clinical diagnostic procedure (Roederer et al., 1997; Sohn and Sautter, 1991).

Aequorin, a compound isolated from a bioluminescent jellyfish *Aequora victoria* has been used extensively in cell biology because it has the unique property of emitting light in the presence of calcium. For example, aequorin has been used to illuminate the calcium wave during sea urchin egg fertilization (Eisen and Reynolds, 1985) a phenomenon described in greater detail in Chapter 5. The photoprotein component of aequorin has been cloned into gene expression vectors, and is currently used to monitor calcium in the cytoplasm and organelles of tissue culture cells (Badminton and Kendall, 1998; Badminton et al., 1995; Brini et al., 1994; Montero et al., 1995; Rutter et al., 1993).

Recently, there has been an exciting new development of another product derived from the bioluminescent jellyfish *A. victoria*, the cloning of green fluorescent protein (GFP) (Chalfie et al., 1994). In jellyfish, GFP absorbs the blue light produced by aequorin and fluoresces green light. GFP has been developed for use as a reporter gene in numerous studies on the regulation of gene expression (Plate XIV). Because GFP fluoresces in living tissues, it is now possible to monitor gene expression continuously, a property of particular value in the study of differentiation in both embryos and tissue culture cells.

There are many other marine products that have contributed to basic and clinical research including enzymes for molecular biology, most notably the Vent DNA polymerase used in the polymerase chain reaction (PCR). PCR, a technique used to amplify minute amounts of DNA or RNA, requires the use of enzymes that are stable at high temperature. A marine microorganism isolated from the deep sea hydrothermal vents yielded the Vent DNA polymerase which is used in high fidelity PCR reactions common to both diagnostic procedures and the gene mapping studies of the Human Genome Project. Marine bacteria have also provided many unique restriction enzymes used in the cloning of DNA. Since we have only begun to investigate marine biodiversity, it is reasonable to expect that marine molecular probes will continue to advance the frontiers of molecular and cellular biology.

THE OCEAN AS A SOURCE OF NEW NUTRITIONAL SUPPLEMENTS

Docosahexenoic acid (DHA) and arachidonic acid (ARA) are the most abundant polyunsaturated fatty acids (PUFAs) in breast milk, and are the predominant structural fatty acids in brain gray matter. High dietary levels of these PUFAs are believed to result in higher levels in the brain, and they have been recommended as nutritional supplements for infants. A marine microalgal species has been discovered that produces large quantities of the fatty-acid docosahexenoic acid (DHA). It is used in an infant formula supplement Formulaid® (Martek Biosciences, Columbia, MD). Marine-derived nutritional supplements, or "nutriceuticals," present a new opportunity for research in the application of marine natural products to human health issues.

CONCLUSIONS

The Discovery and Development of Marine Pharmaceuticals: Needs for the 21st Century

- Marine organisms as a source of pharmaceuticals

The successes to date in the discovery of novel chemicals from marine organisms that have demonstrated potential as new treatments for cancer, infectious diseases, and inflammation, suggest that there needs to be a greater focus on the development of drugs from marine sources. Exploration of unique habitats, such as deep sea environments, and the isolation and culture of marine microorganisms offer two underexplored opportunities for discovery of novel chemicals with therapeutic potential. The successes to date based on a very limited investigation of both deep sea organisms and marine microorganisms suggests a high potential for continued discovery of new drugs. Marine microorganisms are particularly attractive because they fit in with the traditional pharmaceutical "model" of a natural product drug source. Moreover, supply of bulk amounts of a microbially-derived drug can be addressed by large-scale fermentation of bioactive marine microorganisms.

- Expand marine drug discovery beyond cancer to include other diseases

Programs such as the Natural Products National Cancer Drug Discovery Groups (NPNCDDGs) at the National Cancer Institute have been tremendously successful in interfacing non-traditional drug sources, such as marine organisms, with the screening and development potential of major pharmaceutical companies. Similarly, the Small Business Innovative Research (SBIR) grants have fostered interactions on a smaller scale. Other institutes within the NIH should consider developing programs for marine-based drug discovery for diseases that desperately need new therapies, such as neurodegenerative, cardiovascular, and infectious diseases.

In particular, there needs to be a more organized approach to the development of antibiotics from marine sources. The increasingly limited effectiveness of currently available drugs has dire consequences for public health, although the consequences have not yet been felt by the public or the medical community. The United States is faced with the serious threat of re-emerging infectious diseases, such as tuberculosis, indicating that a radical and aggressive approach needs to be taken to control these multiple-drug-resistant pathogens.

- New interdisciplinary research and education programs

Few researchers have training in both medicine and marine science. Cross-training in both disciplines could provide innovative approaches to marine-based biomedical research. For example, marine organisms have already demonstrated their utility as biomedical models, the results of which have been applied to understanding normal and disease processes in humans (cf. Chapter 5). Marine organisms also offer the potential to understand and develop treatments for disease based on the normal physiological role of their secondary metabolites. In some systems, e.g., *Conus* toxins (Hart, 1997; Hopkins et al., 1995), the mechanisms of action of the metabolites are well-known and can be applied to the development of new classes of drugs. In most systems, however, the natural functions of bioactive secondary metabolites are poorly understood. The respective expertise of marine and medical scientists should be optimized and the cross-training of marine and biomedical scientists should be encouraged to study the role of these compounds in nature, to determine how chemical interactions in the ocean can be applied to the development of new drugs, and then to design appropriate bioassays to test their effectiveness against human diseases. One possible way to achieve this goal would be to develop a new graduate student training initiative. The purpose of the program would be to educate graduate students in both marine and medical sciences and to facilitate the development of a strong interface between medicine and marine sciences.

5

Marine Organisms as Models for Biomedical Research

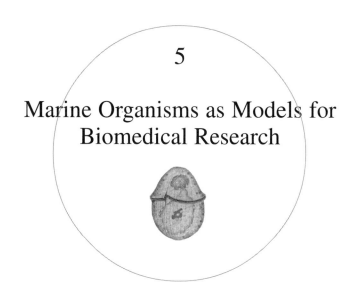

Ought we, for instance, to begin by discussing each separate species...taking each kind in hand independently of the rest, or ought we rather to deal first with the attributes which they have in common in virtue of some common element of their nature, and proceed from this as a basis for the consideration of them separately? (Aristotle, *De partibus animalium*).

Recognition of the conservation of fundamental processes during evolution requires the comparative study of many different species. As indicated by the quotation from Aristotle introducing this chapter, there is a long tradition behind this comparative approach to biology. The analysis of conserved features has been essential to the study of evolution and the reconstruction of evolutionary relationships. Although the comparative approach emerged from the tradition of natural history, it has also been used extensively in the disciplines of physiology, biochemistry, and developmental biology. Some of the insights gleaned from these studies include the thermoregulatory role of countercurrent exchange systems in the circulatory system, the biochemical evolution of proteins through duplication of structural motifs, and the role of cytoplasmic segregation in the development of embryos. Comparative studies have also helped researchers identify which features of organisms are fundamental to function. The assumption is that a conserved feature is so essential for normal biological processes that any modification would be likely to reduce the viability of the organism.

Even after a critical function has been identified, studies using a diversity of organisms have facilitated the elucidation of mechanisms. Determining the best method for investigating a particular biological phenomena frequently requires choosing the animal model that best lends itself to experimentation. For example, the characterization and cloning of the acetylcholine receptor was simplified by the use of the electric organ, a highly modified muscle found in the electric ray *Torpedo*. Nerve impulses are chemically transmitted to muscles through the acetylcholine receptor, a process disrupted in the human neuromuscular disorder, myasthenia gravis. In the ray's specialized electric organ, this receptor protein is found at such a high density in the cell membrane that researchers were able to determine the structure of the protein and clone the gene without first having to purify the protein. This would not have been possible using human or other mammalian tissue. The validity of applying results obtained from the study of a protein in a highly modified fish muscle to normal biochemical processes in human muscle derives from the insight that many fundamental features at the molecular and cellular level are highly conserved even though the evolution of animals has shown dramatic changes in morphological form (Gerhart and Kirschner, 1997).

Studies using marine organisms have had a major influence on biomedical research (Sargent, 1987). This chapter highlights some of the best recognized marine models and elaborates the reasons for their success. However, the first question one might ask is—why marine organisms? At higher taxonomic levels, most biological diversity is found either primarily or exclusively in the ocean. Of 33 modern phyla, only 11 are found in terrestrial habitats while 28 occur in marine habitats. Hence the diversity of life in the sea offers more possibilities for the discovery of organisms for use as models to explore various biological processes. In several of the examples described in this chapter, specific adaptations to the marine environment have been valuable in studying analogous physiological processes in humans. Of particular interest are several marine taxa that share a common origin with mammals. This group, the deuterostomes, includes vertebrates (and other chordates), echinoderms (e.g., sea urchins and sea stars) and tunicates (e.g., sea squirts). Echinoderms appeared early in the fossil record and are the most distant deuterostome relatives of humans. Studies on the differences and similarities of these groups offer insights into the evolution of vertebrates and mammals. Also, many of these organisms have specialized features that have enabled researchers to elucidate complex processes that would be more difficult to study in mammals. A number of examples are listed in Table 5-1, of which some are described in more detail in this chapter.

SEA STARS, SEA URCHINS, TUNICATES, AND SHARKS: THEIR ROLE IN UNDERSTANDING HOW THE BODY FIGHTS INFECTION AND DISEASE

In 1882, Elie Metchnikoff conducted an experiment with sea star larvae (Beck and Habicht, 1996). He punctured a larva with the thorn of a rose and the next day observed tiny motile cells surrounding the thorn. He postulated that the motile cells were a defense mechanism against foreign invaders. The process

TABLE 5-1   Examples of Marine Species Used in Biomedical Research

General Human Health Issue	Taxon	Species (Common Name)	Human Medical Concern
Immunology	Mollusca	*Conus* spp. (cone snails)	Blood disorders, clotting, hemophilia
	Echinodermata	*Arbacia punctulata* (sea urchin)	Cell-mediated immune system responses
	Tunicata	*Botryllus schlosseri* (Golden Star Tunicate, sea squirt)	Immune systems and disorders (self/non-self recognition), AIDS/HIV transmission
	Vertebrata, Fish	*Squalus acanthias* (spiny dogfish shark)	Immune system function, evolution of antibodies, and disease resistance
Neurobiology	Mollusca	*Loligo pealei* (squid)	Neurological studies, behavior
		Aplysia (marine snail)	Nerve impulse transmission
	Arthropoda	*Limulus polyphemus* (horseshoe crab)	Vision Neural basis of behavior
	Vertebrata, Fish	*Opsanus tau* (toadfish)	Balance and equilibrium, nausea
		Squalus acanthias (spiny dogfish shark)	Brain function Vision, glaucoma, cataracts
		Pomacentrus partitus (bicolor damselfish)	Neurofibromatosis
		Electrophorus electricus (electric eel)	Synaptic transmission (NA+ K+ ATPase)
Cell Biology/ Cancer	Mollusca	*Spisula solidissma* (surf clam)	Cell division/cancer
		Loligo pealei (squid)	Cell physiology, intracellular transport, and cellular pH calcium regulation
	Echinodermata	*Arbacia punctulata* (sea urchin)	Cell division/cancer Fertilization and development
	Arthropoda	*Cancer irroratus* (red crab)	Organic ion transport
	Vertebrata, Fish	*Pseudopleuronectes americanus* (winter flounder)	Organic ion transport
Physiology	Arthropoda	*Eriocheir sinensis* (Chinese mitten crab)	Cellular osmoregulation
		Cancer irroratus (red crab)	Detoxification mechanisms
		Carcinus maenas (green shore crab)	Kidney function
	Vertebrata, Fish	*Opsanus tau* (toadfish)	Insulin secretion and diabetes Muscle pathologies
		Squalus acanthias (spiny dogfish shark)	Cystic fibrosis Kidney and heart research
		Pseudopleuronectes americanus (winter flounder)	Detoxification mechanisms
		Anguilla rostrata (American eel)	Kidney function

Metchnikoff observed was phagocytosis. Although phagocytosis had already been observed with human cells, his observations led him to suggest that the process might be a more fundamental defensive mechanism that is widespread in the animal kingdom. Further research showed that echinoderms (e.g., sea urchins and sea stars) possess the features of a basic immune system, one that involves the non-specific action of phagocytic cells (Smith and Davidson, 1994). This is believed to be the oldest form of immunity and, as suspected by Metchnikoff, it appears to be shared by all animals. Metchnikoff's work on sea stars laid the foundation for the disciplines of cellular and comparative immunology and all subsequent studies on the role of cells and phagocytosis in fighting infection and disease in humans.[1]

Further insight into how the human immune system functions has been obtained from studies of other marine deuterostomes. Tunicates, or sea squirts as they are commonly called, have provided a model for studying another aspect of immunity, the ability to distinguish "self" from "non-self." When two sea squirts come into contact they will fuse into one organism if related or grow apart if they are unrelated. Observations of this phenomenon in the field led to the laboratory use of tunicates as a model system for tissue transplantation studies (Raftos, 1994). Tissue recognition or rejection is mediated by the immune system. In sea squirts and humans, similar strategies have been described for determining tissue compatibility involving specialized cells and specific self-recognition molecules.

The recognition of "self" is also important to the reproduction of sea squirts. These animals are hermaphrodites, meaning that the same individual can produce both sperm and eggs. However, there is no fertilization of eggs by sperm produced from the same animal. Studies of the tunicate *Botryllus schlosseri* also showed that the sperm from one individual does not bind to blood cells from the same individual, although they do bind to blood cells from a different individual. This observation prompted some AIDS researchers to conduct a similar experiment with human sperm and blood cells (Scofield, 1997). They found that human sperm, like tunicate sperm, exclusively bind to blood cells from other individuals and discriminate through detection of a self-recognition molecule on the surface of the blood cells (Scofield et al., 1992). This discovery may play a role in understanding how the AIDS virus is transmitted and help biomedical researchers devise protocols for reducing disease transmission.

The phagocytic cells of marine invertebrates constitute the most fundamental type of animal immune system. The response is rapid and is referred to as natural or innate immunity. Vertebrates possess an additional form of immunity referred to as acquired or adaptive immunity. This form of immunity relies on a combinatorial genetic mechanism that generates millions of specific recognition molecules in specialized defense cells, the B and T lymphocytes. Insight into the evolution

---

[1] The importance of Elie Metchnikoff's work was recognized in 1908 when he was awarded the Nobel Prize for Physiology or Medicine.

of this important feature of the human immune system has come from the study of sharks. Sharks and other cartilaginous fishes are the most primitive group of vertebrates with this adaptive, combinatorial immune system. Sharks first appeared in the fossil record between four and five hundred million years ago and their long history suggests that this "new" immune system gave sharks an evolutionary advantage which allowed them to survive while other taxa became extinct (Litman, 1996). The immune system of sharks shares some similarities to the human fetal immune system because the predominant circulating class of antibodies in the shark resembles the earliest produced immunoglobulin M (IgM) macroglobulins of fetal humans. Sharks also possess innate anti-microbial antibodies, T-cell receptors, and major histocompatability antigens (MHCs). Thus, sharks present a comparative model for studying both innate and acquired immunity and autoimmunity, which is the underlying cause of several human diseases such as lupus and rheumatoid arthritis. In addition, sharks possess the steroid squalamine which has an immunomodulatory function and antimicrobial activity with pharmaceutical potential (Moore et al., 1993). Although many aspects of the immune system can be studied in mammals, studies that have taken a comparative approach have provided valuable insight into the basic mechanism of immune responses. The comparative approach may hold the key to developing new therapies for autoimmune and immunodeficiency diseases (Marchalonis and Schluter, 1994).

SEA URCHIN AND CLAM EGGS: THEIR ROLE IN UNDERSTANDING CELL BIOLOGY AND BIOCHEMISTRY

Sea urchins have served as experimental models for more than 100 years. Many species produce tremendous quantities (millions to billions; Plate XV) of large, clear eggs that lack external coatings. An early example of their utility was Otto Warburg's demonstration in 1908 of the increase in oxygen consumption that occurs following fertilization, despite the relative insensitivity of his methods.

The cell cycle is the orderly sequence of events in which a cell first reproduces its genetic material and then divides. In the life of an organism, cell division begins following the fertilization of the egg, defines the early growth and differentiation of the embryo, and continues throughout adulthood, especially in tissues like blood and intestinal mucosa. The cell cycle is closely regulated by a group of proteins, the cyclins (Pines, 1996), which were originally identified in sea urchins (Evans et al., 1983). The key features of sea urchin eggs that made this discovery possible are their abundance and the synchronous division of cells after fertilization. Researchers radiolabeled newly fertilized eggs to tag proteins synthesized during the first few cell divisions. They found that while most proteins accumulate through succeeding cell cycles, one protein, cyclin, is remarkable in that it is synthesized and destroyed once per cell cycle, appearing and disappearing periodically as the cell divides.

Cyclin A was first cloned from the surf clam and its connection to the cell cycle was confirmed by its ability to initiate meiosis in frog eggs and mitosis in somatic cells. Subsequent studies have shown that the synthesis and destruction of cyclins are the key events in the regulation of cell division in all eukaryotic cells, from yeast to human.

Clued by the sequences of the cyclins discovered in marine invertebrates, scientists have identified many different cyclins in mammalian cells. Currently, there is considerable excitement among cancer researchers following the finding that in many human cancers there are mutations that change the function of particular cyclins or the proteins that either regulate or are regulated by cyclin (Hall and Peters, 1996). Thus, the discovery of cyclins in sea urchin eggs revolutionized the study of the mammalian cell cycle and paved the way for new research into the diagnosis and treatment of cancer.

Transient and localized changes in intracellular calcium concentration are ubiquitous signals for many essential cellular responses (Lee, 1997). Sea urchin eggs are the preferred model for investigating calcium signaling because: (1) the eggs are large, transparent, and amenable to microinjection, allowing localized changes in calcium concentration to be readily visualized using dyes and (2) the abundant cytoplasm can be harvested and fractionated to identify the interacting cellular components of the signaling pathway.

More than 50 years ago a spike in intracellular calcium concentration was identified in *Arbacia* eggs following fertilization. This calcium signal triggers formation of a clear proteinaceous fertilization envelope that surrounds the entire egg, and by preventing sperm from reaching the egg membrane, acts as a mechanical block to polyspermy. The calcium signal also contributes to the signal to activate protein and DNA (deoxyribonucleic acid) synthesis at the onset of development. Curiously, the calcium signal occurs as a wave that begins at the site of sperm-egg fusion and sweeps across the entire egg in approximately 30 seconds (Plate XVI). Such waves, which can be repetitive, occur in many different kinds of cells in response to a wide variety of stimuli. The nature of the waves has been elucidated by the discovery that they depend on chemical intracellular messengers, some of which were only recently identified in an exciting series of investigations using sea urchin eggs (Lee, 1997).

The calcium that constitutes these waves is released locally from intracellular stores identified as the endoplasmic reticulum. The initial signal, calcium-induced calcium release, propagates the waves by successively stimulating calcium receptors in adjacent intracellular membranes, resulting in further calcium release. Based on new results using sea urchin eggs, there are at least two separate calcium stores, each with a specific receptor selectively sensitized by one of two novel endogenous chemicals, cyclic adenosine 5'-diphosphate-ribose (cADP-ribose) and nicotinic acid adenine dinucleotide phosphate (NAADP). The cADP-ribose and NAADP were identified by analysis of sea urchin egg cytoplasm.

Additional proteins that modulate cADP-ribose and NAADP activity were also identified in the egg cytoplasm. The importance of these findings is highlighted by the discovery that numerous types of mammalian cells are responsive to cADP-ribose, including neurons and muscle cells in heart, intestine, and skeletal muscle. The cADP-ribose acts on ryanodine receptors that couple increases in intracellular calcium concentration to muscle contraction. Cyclic ADP-ribose also accounts for the calcium-mobilizing action of the nitric oxide signaling pathway.

Thus, the cADP-ribose system that was only recently discovered as a result of continuing investigations of sea urchin eggs appears to be of fundamental importance for mammalian neuromuscular coordination and, with further research, should contribute to advances in the diagnosis and treatment of neuromuscular disorders.

MARINE ORGANISMS: THEIR ROLE IN PHYSIOLOGICAL STUDIES PERTAINING TO FLUID AND ION TRANSPORT, RENAL FUNCTION, AND VOLUME REGULATION

Marine organisms have proven the value of the August Krogh principle, which essentially states that for every problem in physiology, there is one animal ideally suited to solve that problem. In particular, marine invertebrates and fishes have been important as models for osmoregulatory phenomena such as fluid and ion transport, renal function, and volume regulation. It is important to recognize why marine organisms have developed extensive osmoregulatory capabilities in order to appreciate how the principles learned from studying osmoregulation in marine animals have led to an increased understanding of osmoregulatory phenomena in humans. Simply, osmoregulation is necessary for the survival of animals in salt-variable environments ranging from estuaries and mangrove swamps to the Dead Sea. Marine invertebrates and fishes are exposed to salinities as high as 2.5 times that of normal seawater or 2500 mOsm ("milliosmoles/liter," a term used to quantify the total concentration of osmotically active solutes in a solution). A parallel situation occurs in humans; the mammalian extracellular fluid is regulated at about 300 mOsm but some mammalian kidney cells are exposed to concentrations as high as 3000 mOsm. Most marine organisms have permeable tissues that are in direct contact with their environment. Two examples will illustrate the physiological problems that this generates. The first is an osmoconformer, or an organism whose body fluids are the same salt concentration as the seawater that surrounds them. As the salt concentration of the water rises, so does the salt concentration of its body fluids. This causes cellular shrinkage which, if uncorrected, would ultimately result in cell death. Instead, a process called volume regulation takes place. The second example is an osmoregulator, or an organism whose body fluids are maintained at a fixed concentration regardless of the salt concentration of the seawater. A fish or a crab can maintain a blood salt concentration of 300 mOsm. However, gills and portions of the gut are bathed in seawater that is 1000 mOsm. These surfaces are permeable, allowing the entry of salts and the escape of water. This presents two problems: (1) regulating cell volume and (2) restoring body fluid concentration through osmoregulatory mechanisms. In a similar manner, components of both of these processes function in many human organs and, in particular, in the mammalian kidney. Osmoregulation at the intra- and extracellular level is made up of the combined mechanisms of fluid and ion transport and solute or organic osmolyte regulation. The principles learned from studying osmoregulation in marine animals has increased the understanding of how the human kidney maintains the blood at 300 mOsm and also how some kidney cells tolerate the osmotic stress generated by the kidney's role in concentrating urine.

With the exception of the halobacteria, the cells of all organisms have the same adaptive response to a high salt environment. They accumulate intracellular organic osmolytes. It was in the Chinese mitten crab, *Eriocheir sinensis*, that in 1955, (Duchâteau and Florkin, 1955) intracellular accumulation of amino acids with increased salinity was first discovered. Various organisms accumulate solutes intracellularly, but these compounds fall into only a few chemical categories (Yancey et al., 1982). They are polyols, methylamines, and amino acids that help compensate for the high salt environment. This brings up the obvious question of whether there is some highly conserved mechanism that organisms have in common, or whether the similarity in adaptive response is the result of convergent evolution. The hypothesis that there is a conserved mechanism led to the discovery of an osmotic response element (ORE) (Ferraris et al., 1994, 1996) in the flanking region of the aldose reductase (AR) gene, which is responsible for the adaptive accumulation of a type of polyol during high salt stress. The ORE has also been found in other genes and in other species (Ruepp et al., 1996; Takenaka et al., 1994). Hence studies that started with the Chinese mitten crab have led to the discovery of the osmoregulatory function of aldose reductase and subsequently to the finding that inappropriate expression of the same gene in the eye and nerve causes serious damage in diabetic patients. Knowledge of the aldose reductase gene is currently being used to determine the mechanism of inappropriate genetic expression in diabetes.

There are several other examples where studies of properties of marine organisms have helped elucidate the mechanisms underlying a variety of biomedical problems caused by defects in fluid or ion transport. Several of these studies are described below.

The kidney proximal tubules of *Pseudopleuronectes americanus*, the winter flounder, and the urinary bladder of *Cancer irroratus*, the common red crab, have been used as models to examine mechanisms of organic anion transport (David S. Miller, National Institute of Environmental Health Sciences). Organic ion transport is the cellular mechanism by which kidney cells transfer potentially toxic compounds from the blood into the urine. These compounds include drugs, normal and drug metabolites, as well as environmental pollutants and their metabolites.

The gulf toadfish, *Opsanus beta*, has been used for studies on the regulation of nitrogen metabolism and urea excretion. Most fish excrete nitrogen waste as ammonia, a toxic compound that is rapidly diluted in water. However, toadfish have the capability of switching to secreting urea, a much less toxic metabolite. Urea excretion increases as a response of the fish to confinement and crowding (Walsh et al., 1994) and appears to be regulated by mechanisms similar to those regulating urea transport in the mammalian kidney (Wood et al., 1998).

The eye of *Squalus acanthias*, the spiny dogfish shark, is used to understand mechanisms of vision and fluid formation, with relevance for human diseases that affect intraocular pressure (like glaucoma) and lens opacities.

The rectal gland of *Squalus acanthias,* the major salt secreting organ in the shark, has proven to be an ideal model system; atrial natriuretic peptides isolated from the dogfish heart, have been shown to control sodium chloride excretion from the rectal gland in combination with vasopressin (Silva et al., 1996). Atrial natriuretic peptide and vasopressin also regulate salt and water excretion by the human kidney. Further, because of the unusually high density of receptors and channels in the shark rectal gland, it has provided an excellent system for research on the regulation of chloride secretion in higher vertebrates, including humans (Forrest, 1996).

The gills of marine organisms form a key interface between the blood and the environment. The gills of *Carcinus maenas,* the green shore crab have provided a useful model for understanding the regulation of analogous sodium transporters found in the mammalian kidney (Towle et al., 1997). The gills of *Anguilla rostrata*, the American eel, have provided a model system to study how restructuring of the plasma membrane allows kidney cell membranes to adapt when salinity changes (Crockett et al., 1996).

THE TOADFISH: ITS ROLE IN UNRAVELING THE NEURAL CONTROL OF BALANCE AND EQUILIBRIUM

The maintenance of balance and equilibrium in vertebrates is controlled by the vestibular apparatus (the anatomical structures concerned with the vestibular nerve, a somatic sensory branch of the auditory nerve) and its proper functioning is critical for most organisms including humans (Kornhuber, 1974). The vestibular system of toadfish has been used for decades as a model for studying balance and equilibrium. The toadfish was initially chosen for study because it has a broad flat head that makes it relatively easy to study the brain and nerves associated with the vestibular system, and the fish are easy to obtain and adapt readily to the laboratory. This vestibular system is composed of fluid-filled canals lined with small hair cells that sense movement of small crystals, the otoliths. The hairs were first studied in the early 1900s by Cornelia Clapp at the Marine Biological Laboratory in Woods Hole, MA. When the head moves, the otoliths move, and the hair cells send this information to the brain. Vestibular systems developed early in the evolutionary history of vertebrates and did not change greatly as new species evolved. Thus the vestibular system of the toadfish is homologous to the vestibular system in humans and can be used to better understand the basis for human balance disorders.

HORSESHOE CRABS: THEIR ROLE IN UNDERSTANDING RETINAL FUNCTION AND HOW EYES SEE

Knowledge of human vision has its roots in early 20th century studies of the compound eye of the horseshoe crab and the phenomenon of lateral inhibition (Sargent, 1987). The horseshoe crab eye has approximately 1000 photoreceptors whereas the human retina has more than 100 million photoreceptors. Photoreceptors perceive and process visual information about the external environment and then transfer this information to the brain. Hence, studies of photoreceptors and retinas are relevant to understanding the neural basis of behavior. Horseshoe crabs and other marine organisms are particularly amenable to research on retinas because the tissue is readily accessible and can be removed from the animal and studied in the laboratory for long periods of time. The neural network in the horseshoe crab lateral eye is large, there is a cell-based model of this network, and the animal's behavior in the field is well known (Passaglia et al., 1997). Further work on the horseshoe crab may hold the key for deciphering the neurological basis for vision.

APLYSIA: ITS ROLE IN DISCOVERING THE MOLECULAR BASIS OF LEARNING AND MEMORY

The marine snail *Aplysia* has been a valuable model in studies of neurobiology and behavior. It has a simple brain, made up of only 20,000 nerve cells, in contrast to the mammalian brain that comprises around a trillion cells. Moreover, the *Aplysia* nerve cells are large and have characteristic locations, which allow them to be identified in each individual. This allows the comparison of a nerve cell in the experimental animals—or those that have learned a response—with the functionally identical nerve cell in the untrained control animals. Initial studies delineated a simple behavior, the gill-withdrawal reflex, and analyzed its neural circuit (Bailey, C.H. et al., 1996; Kandel et al., 1986). It turned out that even this very simple reflex can be modified by three different elementary forms of learning—habituation, sensitization, and classical conditioning. These forms have similar counterparts in humans. Furthermore, each of these forms of learning gives rise to both short-term memory (lasting minutes) and long-term memory (lasting days to weeks) depending upon the number of training trials. This work provided the first evidence that learning and memory storage involve changes in strength of synaptic connections made between nerve cells of the brain. It also showed that a single synaptic connection can participate in different learning processes and can, in fact, be modified in different ways by these different learning processes (Bailey, C.H. et al., 1996; Kandel et al., 1986). Thus, a single synapse can serve as a site for more than one type of memory storage and these synaptic storage sites have remarkable flexibility.

Subsequent studies elucidated the biochemical mechanisms of memory storage. The investigators began with a short-term form of learning, sensitization, and studied its effects on one readily analyzable component of the gill-withdrawal reflex, namely the synaptic connections between the sensory and motor neurons. They discovered that the transient strengthening of the synaptic connections produced by learning involves an increase in transmitter release. This increase is maintained for the duration of the learned response (memory). The biochemical process underlying memory depends on the recruitment of a major intracellular signaling pathway, the cAMP and cAMP-dependent protein kinase pathway. These studies provided the first biochemical insights into the molecular events for short-term memory storage and illustrated that maintenance of memory is the result of the persistence of the signal transmitted by cAMP (Silva et al., 1998).

The conversion of transient, short-term memory to a self-sustained, long-term memory was also studied in *Aplysia*. The switch from short- to long-term memory begins with the movement to the cell nucleus of the cAMP-dependent protein kinase. Here, this signaling kinase coordinates the activation of a number of genes by turning on CREB-1, an activator of gene expression, and by turning off CREB-2, a repressor of gene expression. With the activation of CREB-1, new genes are transcribed that initiate the growth of new synaptic connections (Silva et al., 1998).

Current studies on mammals confirm that the same principles pertain to their learning and memory. Thus, the ground breaking studies on *Aplysia* have led directly to a molecular understanding of learning and memory and provide an avenue to more effective treatments of cognitive disorders (Bailey, C.H. et al., 1996; Kandel et al., 1986).

THE SQUID GIANT AXON: ITS ROLE IN ESTABLISHING HOW NERVE IMPULSES ARE CONDUCTED

The discovery of the squid giant axon opened up new avenues of neurobiological research (Baker, 1984; Hodgkin, 1958). The axon, which is 0.5 mm in diameter or 1000-fold larger than vertebrate axons, was identified in 1909 by L.W. Williams, who noted that, "the very size of the nerve processes has prevented their discovery, since it is well-nigh impossible to believe that such a large structure can be a nerve fiber." In 1936, J. Z. Young dispelled the critics through his demonstration of an action potential following stimulation of the giant axon with a crystal of sodium citrate.

Before 1939, nerve action potentials were measured indirectly by applying electrodes to the outside of nerves, which gave very limited information. In 1939 Alan Hodgkin and Andrew Huxley began experimenting with squid giant axons at Plymouth.[2] First, Huxley inserted a needle directly into a fiber, intending that they should measure the viscosity of the cytoplasm. When that did not work, Hodgkin and Huxley inserted a glass electrode into the fiber and directly measured the electrical change when a nerve impulse passed. The resting potential was 50 mV negative relative to the surrounding seawater, as expected, but when the fiber was stimulated, the internal potential reached 50 mV positive. Hodgkin and Huxley continued their work on the squid axon, showing that the action potential was propagated by sequential changes in sodium and potassium ion conductance in the membrane throughout the length of the axon. Later investigations showed that nerve conduction is essentially the same in vertebrate neurons as in squid axons. The proteins that form the channels responsible for the sodium and potassium conductance have been cloned and examined in molecular detail, leading to an understanding on an atomic level of the mechanism of conduction. Throughout these studies the squid giant axon remained the preferred system, revealing, for example, the "gating currents" responsible for the changes in conductance that occur during the action potential (Keynes, 1983). Because of their large size, it is relatively easy to insert electrodes of all sorts directly into the neurons without significant damage and it is possible to collect cytoplasm for biochemical analysis. This has led to many other breakthroughs, including pioneering studies in the regulation of intracellular calcium levels and pH.

These revolutionary studies on nerve conduction made possible by the use of this marine model form the foundation for current research in neurophysiology, and the concepts that have emerged are the basis for diagnosis and treatment of disorders of conduction in nerves and other excitable tissues such as heart and skeletal muscle.

USE OF FISH AS MODELS FOR HUMAN DISEASES

Fish offer an alternative to rodents for exploring mechanisms of environmental carcinogenesis. In particular, George Bailey's laboratory at Oregon State University has promoted the use of the rainbow trout as a model for research on compounds that cause cancer. This fish offers the advantages of low rearing costs, an ultrasensitive bioassay using abundant embryos whose small size permits observation of tumor development at minute doses of carcinogen, sensitivity

---

[2] Alan Lloyd Hodgkin and Andrew Fielding Huxley won the Nobel Prize for Physiology or Medicine in 1963 for their work with the giant squid axon.

to many classes of carcinogens, well-described tumor pathology, and responsiveness to tumor promoters and inhibitors (Bailey, G.S. et al., 1996).

The bicolor damselfish, *Pomacentrus partitus*, is the first animal model for one type of human cancer that attacks the nervous system, neurofibromatosis type 1. This cancer is found in natural populations of damselfish and has been demonstrated to be transmissible by injection of extracts of cultured tumor cells. This observation led to the isolation of a retrovirus, possibly the causative agent for damselfish neurofibromatosis (Schmale et al., 1996).

CONCLUSIONS

Although some of the most familiar marine animal models were developed early in this century, the promise of using these organisms in research is rediscovered roughly every decade (NIEHS News, EHP 102:272). The discovery of marine organisms as useful models for biomedical research has frequently been serendipitous, but many successful marine models have emerged from a thorough understanding of the natural history and basic biology of marine organisms. Strategies for supporting biomedical research through the use of marine models include:

• Encourage education and research in natural history, taxonomy, physiology, and biochemistry of marine organisms as the foundation for the development of valuable new models for biomedical research. As science, and biology in particular, grows more and more compartmentalized into sub-disciplines, it becomes increasing important to foster interdisciplinary approaches to biomedical problems through educational and research opportunities.

• Promote the development of nonmammalian models for biomedical research and laboratory culture of targeted marine species. The Comparative Medicine program at the National Center for Research Resources at the NIH currently supports research to explore and develop alternative animal models for biomedical research and funds centers that supply laboratory reared *Aplysia* and various cephalopods (squid, cuttlefish, and octopus). This type of program is important because it provides researchers with a dependable source of experimental animals that can be bred to both reduce individual variability and to develop new genetic strains. Also, collection of marine animals from natural populations in some cases risks the depletion of the species in the wild. Development of culture techniques and facilities for marine organisms will give more alternatives to the use of mammals in research and will promote our understanding of complex biomedical problems.

6

Literature Cited

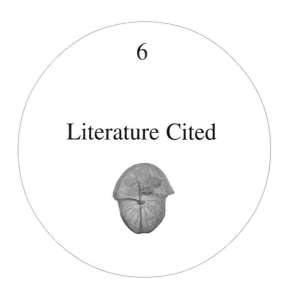

Alamillo, J., and M. Gold. 1998. *Heal the Bay Eighth Annual Beach Report Card.*

Anderson, D.M., S.W. Chisholm, and C.J. Watras. 1983. "Importance of the life cycle events in the population dynamics of *Gonyaulax tamarensis.*" *Mar. Biol.* 76:179-189.

Anderson, D.M. and P.S. Lobel. 1987. "The continuing enigma of Ciguatera." *Biol. Bull.* 172:89-107.

Anderson, D.M. 1989. "Toxic algal blooms and red tides: A global perspective." *Red Tides: Biology, Environmental Science and Toxicology.* T. Okaichi, D.M. Anderson and T. Nemoto, eds. New York, NY: Elsevier Sci. Publ. pp. 11-16.

Anderson, D.M., S.B. Galloway, and J.D. Joseph. 1993. *Marine Biotoxins and Harmful Algae: A National Plan.* Woods Hole Oceanogr. Inst. Tech. Rept., WHOI 93-02. 59 pp.

Anderson, D.M. 1995. "Toxic red tides and harmful algal blooms: A practical challenge in coastal oceanography. *Rev. Geophysics, Suppl. U.S. National Report to the Int. Union of Geodesy and Geophysics 1991-1994.* pp. 1189-1200.

Anune, T. and M. Undestad. 1993. "Diarrhetic shellfish poisoning." *Algal Toxins in Seafood and Drinking Water.* I. Falconer, ed. London: Academic Press. pp. 87-104.

Aristotle. 1982. *De partibus animalium.* Oxford, England: Oxford University Press.

Associated Press (AP). 1998a. "Hurricane Georges' damage reports." *USA Today,* [Online]. Available: http://www.usatoday.com/weather/huricane/1998/wgrgedmg.htm [October 14, 1998].

Associated Press (AP). 1998b. "Oyster Contamination." June 25, 1998.

Baden, D.G., L.E. Fleming, and J.A. Bean. 1995. "Marine toxins." *Handbook of Clinical Neurology* 21:141-175.

Baden, D.G. 1998. Personal communication.

Badminton, M.N., J.M. Kendall, G. Sala-Newby, and A.K. Campbell. 1995. "Nucleoplasmin-targeted aequorin provides evidence for a nuclear calcium barrier." *Exp. Cell. Res.* 216(1):236-243.

Badminton, M.N. and J.M. Kendall. 1998. "*Aequorea victoria* bioluminescence moves into an exciting new era." *Trends Biotechnol.* 16(5):216-224.

Bailey, C.H., D. Bartsch, and E.R. Kandel. 1996. "Toward a molecular definition of long-term memory storage." *Proc. Natl. Acad. Sci. U.S.A.* 93:13445-13452.

Bailey, G.S., D.E. Williams, and J.D. Hendricks. 1996. "Fish models for environmental carcinogenesis: the rainbow trout." *Environ. Health Perspect.* 104 Suppl. 1:5-21.

Baker, P., ed. 1984. *The Squid Axon. Current Topics In Membranes and Transport.* New York: Academic Press. Vol. 22.

Bakun, A. 1973. "Coastal upwelling indices, west coast of North America 1946-71." NOAA Tech. Rep., NMFS SSFR-671, U.S. Department of Commerce.

Barnes, R.D. 1980. *Invertebrate Zoology.* Philadelphia: Saunders College, Holt, Rinehart, and Winston. 4th ed.

Bates, S.S., C.J. Bird, A.S.W. deFreitas, R. Foxall, M. Gilgan, L.A. Hanic, G.A. Johnson, A.W. McCullough, P. Odense, R. Pocklington, M.A. Quilliam, P.G. Sim, J.C. Smith, D.V. Subba Rao, E.C.D. Todd, J.A. Walter, and J.L.C. Wright. 1989. "*Nitzschia pungens* as the primary source of domoic acid, a toxin in shellfish from eastern Prince Edward Island, Canada." *Can. J. Fish. Aquat. Sci.* 46:1203-1215.

Beck, G. and G.S. Habicht. 1996. "Immunity and the invertebrates." *Scientific American.* 275:60-66.

Belofsky, G. N., P.R. Jensen, M.K. Renner, and W. Fenical. 1998. "New cytotoxic sesquiterpenoid nitrobenzoyl esters from a marine isolate of the fungus *Aspergillus versicolor.*" *Tetrahedron.* 54:1715-1724.

Bergmann, W. and R.J. Feeney. 1951. "Contributions to the study of marine products. XXXII. The nucleosides of sponges. I." *J. Org. Chem.* 16:981-987.

Bergmann, W. and D.C. Burke. 1955. "Contributions to the study of marine products. XXXIX. The nucleosides of sponges. III. *Spongothymidine* and *Spongouidine.*" *J. Org. Chem.* 20:1501-1507.

Black, P.G. 1983. "Ocean temperature changes induced by tropical cyclones." Ph.D. Dissertation. State College, PA: The Pennsylvania State University. 278 pp.

Bomber, J.W. and K.E. Aikman. 1988/89. "The Ciguetera dinoflagellates." *Biol. Oceanogr.* 6:291-311.

Bossart, G.D., D.G. Baden, R.Y. Ewing, B. Roberts, and S.D. Wright. 1998. *Toxicologic Pathology.* 26:276-282.

Bouma, M.J. and H.J. van der Kaay. 1996. "The El Niño Southern Oscillation and the historic malaria epidemics on the Indian subcontinent and Sri Lanka: An early warning system for future epidemics?" *Trop. Med. Int. Health.* 1:86-96.

Bouma, M.J. and C. Dye. 1997. "Cycles of malaria associated with El Niño in Venezuela." *JAMA.* 21:1772-1774.

Bouma M.J., R.S. Kovats, S.A. Goubet, J.H. Cox, and A. Haines. 1997a. "Global assessment of El Nino's disaster burden." *Lancet.* 350:1435-1438.

Bouma, M.J., G. Poveda, W. Rojas, D. Chavasse, M. Quinones, J. Cox, and J. Patz. 1997b. "Predicting high-risk years for malaria in Colombia using parameters of El Niño Southern Oscillation." *Trop. Med. Int. Health.* 12:1122-1127.

Brasher, C.W., A. DePaola, D.D. Jones, and A.K. Bej. 1998. "Detection of microbial pathogens in shellfish with multiplex PCR." *Current Microbiology.* 37:101-107.

Brini, M., L. Pasti, C. Bastianutto, M. Murgia, T. Pozzan, and R. Rizzuto. 1994. "Targeting of Aequorin for Calcium Monitoring in Intracellular Compartments." *J. Biolumin. Chemilumin.* 9(3):177-184.

Broecker, W.S. 1997. "Thermohaline circulation, the Achilles Heel of our climate system: Will man-made CO_2 upset the current balance?" *Science.* 278(5343):1582-1588.

Brow, Barbara J. (UNITAR) 1979. "Disaster preparedness and the UN. Advance planning for disaster relief." Pergana Policy Studies, #34.

Bubb, M.R., I. Spector, A.D. Bershadsky, and E.D. Korn. 1995. "Swinholide A is a microfilament disrupting marine toxin that stabilizes actin dimers and severs actin filaments." *J. Biol. Chem.* 270(8):3463-3466.

Buck, E., C. Copeland, J. Zinn, and D. Vogt. 1997. "*Pfiesteria* and related harmful blooms: Natural resource and human health concerns." Congressional Research Service Report for Congress, Report 97-1047ENR. Washington, DC. pp. 1-22

Burkholder, J., H. Glasgow, C. Hobbs, and E. Noga. 1993. *Albermarle-Pamlico Estuarine Study*, Report 93-08. Raleigh, NC: U.S. EPA National Estuary Program and UNC Water Resources Research Institute. pp. 1-58.

Cabelli, V.J., A.P. Dufour, L.J. McCabe, and M.A. Levin. 1982. "Swimming-associated gastroenteritis and water quality." *American Journal of Epidemiology.* 115:606-616.

Carmichael, W., C.L.A. Jones, N.A. Mahmood, and W.C. Theiss. 1986. "Algal toxins and water-based diseases." *CRC Crit. Rev. Environ. Control.* 15:275-993.

Carmichael, W.W. 1992. *A Status Report on Planktonic Cyanobacteria (Blue-Green Algae) and Their Toxins.* U.S. Environmental Protection Agency (EPA)/60/R-92/079:141.

Centers for Disease Control and Prevention (CDC). 1998a. "Water park & *E. coli.*" *Morbid. Mortal. Weekly Rep.* 47(SS-5):1-34

Centers for Disease Control and Prevention (CDC). 1998b. "Outbreak of *Vibrio parahaemolyticus* infections associated with eating raw oysters—Pacific Northwest, 1997." *Morbid. Mortal. Weekly Rep.* 47(22):457-462.

Chalfie, Martin, et. al. 1994. "Green flourescent protein as a marker for gene expression." *Science.* 263:802-805.

Chen, Lincoln C., ed. 1973. *Disaster in Bangladesh-Health Crisis in a Developing Nation.* Oxford, England: Oxford University Press. pp. 119-132.

Cheng, X.C., M. Varoglu, L. Abrell, P. Crews, et. al. 1994. "Chloriolins A-C, chlorinated sesquiterpenes produced by fungal cultures separated from a *Jaspis* marine sponge." *J. Org. Chem.* 59:6344.

Colwell, Rita R. 1996. "Global climate and infectious disease: The cholera paradigm." *Science.* 274:2025-2031.

Copley, Jon. "Recipe for disaster." *New Scientist.* November 14, 1998.

Crockett, E.L., S.M. Vekasi, and E.E. Wilkes. 1996. "Metabolic fuel preferences in gill and liver tissues from freshwater- and seawater-acclimated eel (*Anguilla rostrata*)." *The Bulletin of the Mount Desert Island Biological Laboratory.* 35: 83-84.

Czachor, J.S. 1992. "Unusual aspects of bacterial water-borne illnesses." *Am. Fam. Physician.* 46(3):797-804.

Davidson, B.S. 1995. "New dimensions in natural products research: Cultured marine microorganisms." *Curr. Opin. Biotechnol.* 6:284-291.

Deardorff, T.L. and R.M. Overstreet. 1991. "Seafood-transmitted zoonoses in the United States: The fishes, the dishes, and the worms." *Microbiology of Marine Food Products.* D.R. Ward and C. Hackney, eds. New York, NY: Van Nostrand Reinhold. pp. 211-265.

Delaney, J.E. 1985. *Laboratory Procedures for the Examination of Seawater & Shellfish.* A.E. Greenbery and D.A. Hunt, eds. Washington, DC: Am. Publ. Health Assoc. pp. 64-80.

Dickson, R.R., J.R.N. Lazier, J. Meincke, P.B. Rhines and J. Swift. 1996. "Long-term coordinated change in the convective activity of the North Atlantic." *Prog. Oceanography.* 38: 241-295.

Dilley, M. and B.N. Heyman. 1995. "ENSO and disaster: droughts, floods and El Niño/Southern Oscillation warm events." *Disasters.* 3:181-193

Duchâteau, G. and M. Florkin. 1955. "Concentration du milieu extérieur et état stationnaire du pool des acides aminés non protéiques des muscles d' *Eriocheir sinensis,* Milne Edwards." *Arch. Int. Physiol. Biochim.* 63:249-251.

Dufour, A.P. 1984. "Health effects criteria for fresh recreational waters." U.S. Environmental Protection Agency (EPA)-600/1-884-004.

Ecology and Oceanography of Harmful Algal Blooms (ECOHAB). 1995. *The Ecology and Oceanography of Harmful Algal Blooms: A National Research Agenda.* Woods Hole, Massachusetts: Woods Hole Oceanographic Institution. 66 pp.

Eisen, A. and G.T.Reynolds. 1985. "Source and sinks for the calcium released during fertilization of single sea urchin eggs." *J. Cell Biol.* 100(5):1522-1527.

Elsberry, R.L., T.S. Friam, and R.N.J. Trapnell. 1976. "A mixed layer model of the oceanic thermal response to hurricanes." *J. Geophysical Res.* 81:1153-1162.

Elsberry, R.L., G.J. Holland, H. Gerrish, M. DeMaria, and C.P. Guard. 1992. "Is there any hope for tropical cyclone intensity prediction? A panel discussion." *Bull. Am. Meteor. Soc.* 73:264-275.

Elsberry, R.L. 1995. "Tropical cyclone motion." *Global Perspectives of Tropical Cyclones.* Russell Elsberry, ed. World Meterological Organization Report TCP-38. Geneva, Switzerland. pp. 106-107.

Emanuel, K., D. Raymond, A. Betts, L. Bosart, C. Bretherton, K. Droegemeir, B. Farrell, J.M. Fritsch, R. Houze, M. LeMone, D. Lilly, R. Rotunno, M. Shapiro, R. Smith, and A. Thorpe. 1995. "Report of the first prospectus development team of the US Weather Research Program to NOAA and the NSF." *Bull. Am. Met. Soc.* 76:1194-1208.

Emerson, D.J. and V.J. Cabelli. 1982. "Extraction of *Clostridium perfringens* spores from bottom sediment samples." *Applied and Environmental Microbiology.* 44:1144-1149.

Evans, T., E.T. Rosenthal, J. Youngblom, D. Distel, and T. Hunt. 1983. "Cyclin: a protein specified by maternal mRNA in sea urchin eggs that is destroyed at each cleavage division." *Cell.* 33:389-396.

Fenical, W. 1993. "Marine bacteria: Developing a new chemical resource." *Chem. Rev.* 93:1673-1683.

Ferguson, J.A., B.G. Healey, K.S. Bronk, S.M. Barnard, and D.R. Walt. 1997. "Simultaneous monitoring of pH, CO_2, and O_2 using an optical imaging fiber." *Analytica Chimica Acta.* 340:123-131.

Ferraris, J.D., C.K. Williams, B.M. Martin, M.B. Burg, and A. Garcia-Perez. 1994. "Cloning, genomic organization, and osmotic response of the aldose reductase gene." *Proc. Nat. Acad. Sci. U.S.A.* 91:10742-10746.

Ferraris, J.D., C.K. Williams, K-Y, Jung, J.J. Bedford, M.B. Burg, and A. Garcia-Perez. 1996. "ORE, a eukaryotic minimal essential osmotic response element." *J. Biol. Chem.* 271:18318-18321.

Forrest, J.N., Jr. 1996. "Cellular and molecular biology of chloride secretion in the shark rectal gland: regulation by adenosine receptors." *Kidney International.* 49:1557-1562.

Fraga, S., D.M. Anderson, I. Bravo, B. Reguera, K.A. Steidinger, and C.M. Yentsch. 1988. "Influence of upwelling relaxation on dinoflagellates and shellfish toxicity in Ria de Vigo, Spain." *Est. Coast and Shelf Sci.* 27:349-361.

Franks, P.J.S. and D.M. Anderson. 1992a. "Alongshore transport of a toxic phytoplankton bloom in a buoyancy current: *Alexandrium tamarense* in the Gulf of Maine." *Marine Biology.* 112:153-164.

Franks, P.J.S. and D.M. Anderson. 1992b. "Toxic phytoplankton blooms in the southwestern Gulf of Maine: testing hypotheses of physical control using historical data." *Marine Biology.* 112:165-174.

Freudenthal, A.R. 1990. "Public health aspects of Ciguatera poisoning contracted on tropical vacation by North American tourists." *Toxic Marine Phytoplankton.* E. Graneli, B. Sundstron, L. Edler, and D.M. Anderson, eds. New York, NY: Elsevier Sci. Publ. pp. 463-468.

Garcon, V.C., K.D. Stolzenbach, and D.M. Anderson. 1986. "Tidal flushing of an estuarine embayment subject to recurrent dinoflagellate blooms." *Estuaries.* 9(3):179-187.

Garrett, E.S., M.L. Jahncke, and J.M. Tennyson. 1997. "Microbiological hazards and emerging food-safety issues associated with seafoods." *Journal of Food Protection.* 60:1409-1415.

General Accounting Office (GAO). 1984. GAO/HRD-84-36, July 14, 1984.

Geraci, J.R., D.M. Anderson, R.J. Timperi, D.J. St. Aubin, G.A. Early, J.H. Perscott and C.A. Mayo. 1989. "Humpback whales (*Megaptera novaeangliae*) fatally poisoned by dinoflagellate toxin." *Can. J. Fish. Aquat. Sci.* 46:1895-1898.

Gerhart, J. and M. Kirschner. 1997. *Cells, Embryos, and Evolution: Toward a Cellular and Developmental Understanding of Phenotypic Variation and Evolutionary Adaptability.* Malden, Massachusetts: Blackwell Science, Inc. pp. 1-44.

Glaser, K.B. and R.J. Jacobs. 1986. "Molecular pharmacology of manoalide. Inactivation of bee venom phospholipase A2." *Biochem. Pharmacol.* 35:449-453.

Glazer, A.N. 1988. "Phycobiliproteins." *Methods Enzymol.* 167:291-303.

Glazer, A.N. 1989. "Light guides. Directional energy transfer in a photosynthetic antenna." *J. Biol. Chem.* 264(1):1-4.

Goyal, S.M., C.P. Gerba, and J.L. Melnick. 1979. "Human enteroviruses in oysters and their overlying waters." *Appl. Environ. Microbiol.* 37(3):572-581.

Grattan, L.M., D. Oldach, T.M. Perl, M.H. Lowitt, D.L. Matuszak, C. Dickson, C. Parrott, R.C. Shoemaker, C.L. Kauffman, M.P. Wasserman, J.R. Hebel, P. Charache, and J.G. Morris, Jr. 1998. "Learning and memory difficulties after environmental exposure to waterways containing toxin-producing *Pfiesteria* or *Pfiesteria*-like dinoflagellates." *Lancet.* 352(9127):532-539.

Gray, W.M. 1984. "Atlantic seasonal hurricane frequency: Part I. El Niño and 30mb quasi-biennial oscillaton influences." *Mon. Wea. Rev.* 112:1649-1668.

Gray, W.M., C.W. Landsea, P.W. Mielke, Jr., and K.J. Berry. 1993. "Predicting Atlantic basin seasonal tropical cyclone activity by 1 August." *Wea. Forcasting.* 8:73-86.

Gray, W.M., C.W. Landsea, P.W. Mielke, Jr., and K.J. Berry. 1994. "Predicting Atlantic basin seasonal tropical cyclone activity by 1 June." *Wea. Forcasting.* 9:103-115.

Grimes, D.J., F.L. Singleton, and R.R. Colwell. 1984. "Allogenic succession of marine bacterial communities in response to pharmaceutical waste." *Journal of Applied Bacteriology.* 57:247-261.

Grimes, D.J., R.W. Attwell, P.R. Brayton, L.M. Palmer, D.M. Rollins, D.B. Roszak, F.L. Singleton, M.L. Tamplin, and R.R. Colwell. 1986. "Fate of enteric pathogenic bacteria in estuarine and marine environments." *Microbiological Sciences.* 3:324-329.

Grimes, D.J., and R.R. Colwell. 1986. "Viability and virulence of *Escherichia coli* suspended by membrane chamber in semitropical ocean water." *FEMS Microbiology Letters.* 34:161-165.

Grimes, D.J. 1991. "Ecology of estuarine bacteria capable of causing human disease: A Review." *Estuaries.* 14:345-360.

Gubler, D.J. 1998. "Dengue and Dengue Hemorrhagic Fever." *Clin. Microbiology. Reviews.* 11(3):480-496.

Gunasekera, S.P., M. Gunasekera, R.E. Longley, and G. Schulte. 1990. "Discodermolide: A new bioactive polyhydroxylated lactone from the marine sponge *Discodermia dissoluta.*" *J. Org. Chem.* 55:4912-4915.

Guo, C. and P.A. Tester. 1994. "Toxic effect of the bloom-forming *Trichodesmium* sp. (Cyanophyta) to the Copepod *Acartia tonsa.*" *Natural Toxins.* 2:222-227.

Gustafson, K., M. Roman, and W. Fenical. 1989. "The macrolactins, a novel class of anti-viral and cytotoxic macrolides from a deep sea marine bacterium." *J. Amer. Chem. Soc.* 111:7519-7524.

Hackney, C.R., M.B. Kilgen, and H. Kator. 1992. "Public health aspects of transferring mollusks." *Journal of Shellfish Research.* 11: 45-57.

Hahn, R.A., E.D. Eaker, N.D. Barker, S.M. Teutsch, W.A. Sosniak, and N. Krieger. 1996. "Poverty and death in the U.S." *International Journal of Health.* 26(4):673-690.

Haines, A. and M. Parry. 1993. "Climate change and human health." *Journal of the Royal Sociey of Medicine.* 86(Dec.):707-711.

Hall, M. and G. Peters. 1996. "Genetic alterations of cyclins, cyclin-dependent kinases and Cdk inhibitors in human cancer." *Adv. Cancer Res.* 68:67-108.

Hallegraeff, G.M. 1993. "A review of harmful algal blooms and their apparent global increase." *Phycologia.* 32:79-99.

Hart, S. 1997. "Cone snail toxins take off. Potent neurotoxins stop fish in their tracks and may provide new pain therapies." *Bioscience.* 47(3):131-134.

Hawser, S.P., G.A. Gold, D.G. Capone, and E.J. Carpenter. 1991. "A neurotoxic factor associated with the bloom-forming cyanobacterium *Trichodesmium.*" *Toxicon.* 29:277-278.

Helfrich, P., A.H. Banner, and Bernice P. Bishop. 1968. *Mus. Occas. Pap.* 23:371-382.

Henderson-Sellers, A., H. Zhang, G. Berz, K. Emanuel, W. Gray, C. Landsea, G. Holland, J. Lighthill, S.-L.Shieh, P. Webster, and K. McGuffie. 1998. "Tropical cyclones and global climate change: A post-IPCC assessment." *Bull. Am. Met. Soc.* 79(1):19-38.

Hirata, Y. and D. Uemura. 1986. "Halichondrins—Antitumor polyether macrolides form a marine sponge." *Pure Appl. Chem.* 58:701-710.

Hodgkin, A. L. 1958. "Ionic movements and electrical activity in giant nerve fibers." *Proc. R. Soc. Lond. B.* 148:1-37.

Hopkins, C., M. Grilley, C. Miller, Ki-Joon Shon, L.J. Curz, W.R. Gray, J. Kykert, J. Rivier, D. Yoshikami, and B.M. Olivera. 1995. "A new family of *Conus* peptides targeted to the nicotinic acetylcholine receptor." *J. Biol. Chem.* 270(38):22361-22367.

Howard, B.J., J.F. Keiser, T.F. Smith, A.S. Weissfeld, and R.C. Tilton. 1994. *Clinical and Pathogenic Microbiology.* Mosby: St. Louis.

Howard Hughes Medical Institute (HHMI). 1998. "Sponge toxins halts molecular motor." [Online]. Available: http://www.hhmi.org/news/goldstein.htm [Sept. 21, 1998].

Huq, A., R.R. Colwell, R. Rahaman, A. Ali, M.A. Chowdhury, S. Parveen, D.A. Sack, and E. Russek-Cohen. 1990. "Detection of *Vibrio cholerae* O1 in the aquatic environment by fluorescent-monoclonal antibody and culture methods." *Applied and Environmental Microbiology.* 56:2370-2373.

Hurrell, J.W. 1995. "Decadal trends in the North Atlantic Oscillation: Regional temperatures and precipitation." *Science.* 269:676-679.

Infectious Agents Surveillance Report (IASR). 1996. "*Vibrio parahaemolyticus,* Japan, 1994-1995." 17(7):No. 197. [Online]. Available: www.nih.gov.jp/niaid.index.html [January 6, 1999].

Institute of Medicine (IOM). 1991. *Seafood Safety.* Washington, D.C.: National Academy Press. 452 pp.

International Centre for Diarrhoeal Disease Research, Bangladesh (ICDDR,B). 1998. "ICDDR,B Strategic Plan—To the Year 2000—Executive Summary." [Online]. Available: http:// www.icddrb.org.sg/straexec.htm [1998, July 21]

International Labor Organization (ILO)-World Health Organization. 1984. *Aquatic (Marine and Freshwater) Biotoxins. Environmental Health Criteria.* Geneva, Switzerland: World Health Organization, 37.

Ireland, C.M., B.R. Copp, M.P. Foster, L.A. McDonald, D.C. Radisky, and J.C. Swersey. 1993. *Marine Biotechnology, Vol. 1: Pharmaceutical and Bioactive Natural Products.* D.H. Attaway and O.R. Zaborsky, eds. New York, NY: Plenum Press. pp. 1-43.

Jacob, D.S., L.K. Shay, and A.J. Mariano. 1996. "The mixed layer heat balance during Hurricane Gilbert." Eighth Conference on Air-Sea Interactions and Symposium on GOALS. Boston, MA: American Meteorological Society. pp. 23-27.

Jacob, D.S., L.K. Shay, A.J. Mariano, and P.J. Black. 1998. "The 3-dimensional mixed layer heat balance during Hurricane Gilbert." *J. Phys. Oceanogr.* (submitted).

Jeglitsch, G., K. Rein, D.G. Baden, and D.J. Adams. 1998. "Brevetoxin-3 (PbTx-3) and its derivatives modulate single tetradotoxin-sensitive sodium channels in rat sensory neurons." *J. Pharmacol. Exp. Ther.* 284(2):516-25.

Jetten, T.H. and D.A. Focks. 1997. "Changes in the distribution of dengue transmission under climate warming scenarios." *Am. J. Trop. Med. Hyg.* 57(3):285-97.

Kakeya, H., I. Takahashi, G. Okada, K. Isono, et.al. 1995. "Epolactaene, a novel neuritogenic compound human neuroblastoma cells, produced by a marine fungus." *J. Antibiot.* 48:733-735.

Kalechman, Y., M. Albeck, and B. Sredni. 1992. "In vivo synergistic effect of the immunomodulator AS101 and the PKC inducer bryostatin." *Cell Immunol.* 143(1):143-153.

Kandel, E.R., M. Klein, V.F. Castellucci, S. Schacher, and P. Goelet. 1986. "Some principles emerging from the study of short- and long-term memory." *Neurosci. Res.* 3:489-520.

Kaneko, T., and R.R. Colwell. 1973. "Ecology of *Vibrio parahaemolyticus* in Chesapeake Bay." *Journal of Bacteriology.* 113:24-32.

Keynes, R. D. 1983. "Voltage-gated ion channels in the nerve membrane." The Croonian Lecture. *Proc. R. Soc. Lond. B.* 220:1-30.

Kobayashi, J. and M. Ishibashi. 1993. "Bioactive metabolites of symbiotic marine microorganisms." *Chem. Rev.* 93:1753-1769.

Koh, E.G., J.H. Huyn, and P.A. LaRock. 1994. "Pertinence of indicator organisms and sampling variables to *Vibrio* concentrations." *Appl. Environ. Microbiol.* 60(10):3897-3900.

Kornhuber, H.H., ed. 1974. "Vestibular system Part 1: Basic mechanisms." *Handbook of Sensory Physiology.* Berlin, Germany: Springer Verlag.

LaFranchi, H. 1998. "A flood of ideas to build Central America anew." *Christian Science Monitor.* December 4, 1998.

Lamb, P. and R.A. Peppler. 1987. "North Atlantic Oscillation: Concept and application." *Bull. Am. Meterol. Soc.* 68(10):1218-1225.

Landsea, C.W., W.M. Gray, P.W. Mielke, Jr., and K.J. Berry. 1994. "Seasonal forecasting of Atlantic hurricane activity." *Weather.* 49:273-284.

Lee, H.C. 1997. "Mechanisms of calcium signaling by cyclic ADP-ribose and NAADP." *Physiol. Rev.* 77:1133-1164.

Lin, Y.Y. M.A. Risk, S.M. Ray, D. VanEngen, J. Clardy, J. Golick, J.C. James, and K. Nakanishi. 1981. *J. Am. Chem. Soc.* 03:6773-6775.

Lipp, E.K. and J.B. Rose. 1997. "The role of seafood in foodborne diseases in the United States of America." *Rev. Sci. Tech.* 16(2):620-640.

Litaudon, M., J.B. Hart, J.W. Blunt, R.J. Lake, et al. 1994. "Isohomohalichondrin B, a new antitumour polyether macrolide from the New Zealand deep-water sponge *Lissodendoryx* Sp." *Tetrahedron Lett.* 35:9435-9438.

Litman, G. 1996. "Sharks and the origins of vertebrate immunity." *Scientific American.* 275:67-71.

Longley, R.E., D. Caddigan, D. Harmody, M. Gunasekera, and S.P. Gunasekera. 1991a. "Discodermolide—A new, marine-derived immunosuppressive compound. 1. In vitro studies." *Transplantation.* 52:650-656.

Longley, R.E., D. Caddigan, D. Harmody, M. Gunasekera, et. al. 1991b. "Discodermolide—A new, marine-derived immunosuppressive compound. 2. In vivo studies." *Transplantation* 52: 656-661.

Look , S.A., W. Fenical, R.S. Jacobs, and J. Clardy. 1986. "The pseudopterosins: Anti-inflammatory and analgesic natural products from sea whip *Pseudopterogorgia elisabethae*." *Proc. Natl. Acad. Sci. U.S.A.* 83:6238-6240.

Manabe, S. and R.J. Stouffer. 1994. "Multiple-century response of a coupled ocean-atmosphere model to an increase of atmospheric carbon dioxide." *J. Climate.* 7:5-23.

Mann, M.E., R.S. Bradley, and M.K. Hughes. 1998. "Global-scale temperature patterns and climate forcing over the past six centuries." *Nature.* 392:779-787.

Marchalonis, J.J. and S.F. Schluter. 1994. "Development of an immune system." *Primordial Immunity: Foundations for the vertebrate immune system.* G. Beck, G.S. Habicht, E. L. Cooper, and J. J. Marchanlonis, eds. *Ann. N. Y. Acad. Sci.* 712:1-12.

Marks, F.D. and L.K. Shay and the PDT. 1998. "Landfalling tropical cyclones: Forecast problems and associated research opportunities." *Bull. Am. Meterol. Soc.* 79(2):305-323.

Martin, J.L., K. Haya, and D.J. Wildish. 1993. "Distribution and domoic acid content of *Nitzschia pseudodelicatissima* in the Bay of Fundy." *Toxic Phytoplankton Blooms in the Sea.* T.J. Smayda and Y. Shimizu, eds. Amsterdam: Elsevier Sci. Publ. B.V. pp. 613-618.

Martyr, P. and De Orbo Novo. 1912. *The Eight Decades of Peter Martyr.* F.A. MacNutt, transl. New York, NY: G.A. Putnam Sons. Vol. 2.

Mason, J. and P. Cavalie. 1965. "Malaria epidemic in Haiti following a hurricane." *Am. J. Trop. Med. Hyg.* 14:533-539.

Mata, L. 1994. "Cholera El Tor in Latin America, 1991-1993." *Ann. N. Y. Acad. Sci.* 740:55-68.

Matthews, J.B., J.A. Smith, and B.J. Hrnjez. 1997. "Effects of F-actin stabilization or disassembly on epithelial Cl- secretion and Na-K-2Cl cotransport." *Am. J. Physiol.* 272 (1 Pt 1):C254-C262.

McCarthy, S.A. and J.L. Gaines. 1992. "Toxigenic *Vibrio cholerae* 01 and cargo ships entering Gulf of Mexico." *Lancet.* 339:624-625.

McCarthy, S.A. and F.M. Khambaty. 1994. "International dissemination of epidemic *Vibrio cholerae* by cargo ship ballast and other nonpotable waters." *Applied and Environmental Microbiology.* 60:2597-2601.

McCarthy, S.A. 1996. "Effects of temperature and salinity on survival of toxigenic *Vibrio cholerae* 01 in seawater." *Microbial Ecology.* 31:167-175.

McConnell, O.J., R.E. Longley, and F.E. Koehn. 1994. *The Discovery of Natural Products with Therapeutic Potential.* Gullo, V.P., ed. Boston, Mass: Butterworth-Heinemann. pp. 109-174.

McIntyre, Sally. 1997. "The Black Report and beyond." *Social Science and Medicine.* 44(6):723-746.

Mendez, S. 1992. "Update from Uruguay." *Harmful Algae News.* No. 63:5.

Metcalf, T.G., J.L. Melnick, and M.K. Estes. 1995. "Environmental virology: From detection of viruses in sewage and water by isolation to identification by molecular biology – a trip over 50 years." *Annual Review of Microbiology.* 49:461-487.

Middlebrooks, B.L. 1993. "Immunoclassification of wastewater particulates in shellfish growing waters." National Indicator Study Peer Review Workshop, Baltimore, MD.

Montero, M., M. Brini, R. Marsault, J. Alvarez, R. Sitia, T. Pozzan, and R. Rizzuto. 1995. "Monitoring dynamic changes in free Ca2+ concentration in the endoplasmic reticulum of intact cells." *EMBO J.* 14(22):5467-5475.

Moore, K.S., S. Wehrli, H. Roder, M. Rogers, J.N. Forrest Jr., D. McCrimmon, and M. Zasloff. 1993. "Squalamine: an aminosterol antibiotic from the shark." *Proc. Natl. Acad. Sci. U.S.A.* 90:1354-1358.

Morse, D.E. 1991. "Molecular signals, receptors and genes controlling reproduction, development and growth: Practical applications for improvements in molluscan aquaculture." *Bulletin of the Institute of Zoology, Academia Sinica.* 16:441-454.

Motes, M.L., A. DePaola, D.W. Cook, J.E. Veazey, J.C. Hunsucker, W.E. Garthright, R.J. Blodgett, and S.J. Chirtel. 1998. "Influence of water temperature and salinity on *Vibrio vulnificus* in Northern Gulf and Atlantic Coast oysters." *Applied and Environmental Microbiology.* 64:1459-1465.

Murty, T.S., R.A. Flather, and R.F. Henry. 1986. "The storm surge problem in the Bay of Bengal." *Prog. Oceanogr.* 16:195-233.

National Marine Fisheries Service (NMFS). 1998. "Foreign trade in fisheries." [Online]. Available: http://www.st.nmfs.gov/st1/trade/index.html [July 21, 1998].

National Oceanic and Atmospheric Administration (NOAA). 1985. *National Estuarine Inventory: Data Atlas, Volume 1: Physical and Hydrologic Characteristics.* Rockville, Md.: Strategic Assessment Branch, Ocean Assessments Division. 104 pp.

National Oceanic and Atmospheric Administraton (NOAA). 1996. "Hurricane Opal: Service Assessment Team Report." Silver Spring, MD: U.S. Department of Commerce, NOAA, National Weather Service. 76 pp.

National Oceanic and Atmospheric Administration (NOAA). 1997a. *NOAA's Estuarine Eutrophication Survey. Volume 2: Mid-Atlantic Region.* Silver Spring, Md.: Office of Resources Conservation and Assessment. 51 pp.

National Oceanic and Atmospheric Administration (NOAA). 1997b. *NOAA's Estuarine Eutrophication Survey. Volume 4: Gulf of Mexico Region.* Silver Spring, Md.: Office of Resources Conservation and Assessment. 77 pp.

National Oceanic and Atmospheric Administration (NOAA). 1998a. *NOAA's Estuarine Eutrophication Survey. Volume 5: Pacific Coast Region.* Silver Spring, Md.: Office of Resources Conservation and Assessment. 75 pp.

National Oceanic and Atmospheric Administration (NOAA). 1998b. "Jan.-Sept. global surface mean temperature anomalies." [Online]. Available: http://www.ncdc.noaa.gov/ol/climate/research/1998/sep/j-striad_Pg.gif [October 30, 1998].

National Oceanic and Atmospheric Administration (NOAA). 1998c. Year of the Ocean Discussion Papers. Office of the Chief Scientist, NOAA, Washington, DC.

National Research Council (NRC). 1993. *Applications of Analytical Chemistry to Oceanic Carbon Cycles.* Washington, D.C.: National Academy Press. pp 33-71.

National Research Council (NRC). 1994a. *Molecular Biology in Marine Science.* Washington, D.C.: National Academy Press. 76 pp.

National Research Council (NRC). 1994b. *Priorities for Coastal Ecosystem Science.* Washington, D.C.: National Academy Press. 106 pp.

National Research Council (NRC). 1996. *Stemming the Tide: Controlling Introductions of Nonindigenous Species by Ships' Ballast Water.* Washington, D.C.: National Academy Press. 160 pp.

National Research Council (NRC). 1998a. *Decade-to-Century-Scale Climate Variability and Change.* Washington, D.C.: National Academy Press. 142 pp.

National Research Council (NRC). 1998b. *Global Environmental Change: Research Pathways for the Next Decade, Overview.* Washington, D.C.: National Academy Press. 82 pp.

Natural Resources Defense Council (NRDC). 1998. "Testing the waters—1998: Has your vacation beach cleaned up its act?" [Online]. Available: http://www.igc.org/nrdc/nrdcpro/ttw/titinx.html [1998, September 8].

Nicholls, N., G.V. Gruza, J. Jouzel, T.R. Karl, L.A. Ogallo, and D.E. Parker. 1995. "Observed climate variability and change." *Climate Change 1995, The Science of Climate Change.* Contribution of Working Group I to the Second Assessment Report of the Intergovernmental Panel on Climate Change. J.T. Houghton, L.G. Meira Filho, B.A. Callander, N. Harris, A. Kattenberg, and K. Maskell, eds. New York, NY: Cambridge University Press. 570 pp.

Numata, A., C. Takahashi, T. Matsushita, T. Miyamoto, K. Kawai, Y. Usami, E. Matsumura, M. Inoue, H. Ohishi, and T. Shingu. 1992. "Fumiquinazolines, novel metabolites of a fungus isolated from a saltfish." *Tetrahedron Lett.* 33:1621-1624.

Office of the Federal Coordinator for Meteorological Services (OFCM). 1997. *National Plan for Tropical Cyclone Research and Reconnaissance (1997-2002).* Office of the Federal Coordinator for Meteorological Services and Supporting Research, FCM-P25-1997. 112 pp.

Okami, Y. J. 1993. "The search for bioactive metabolites form marine bacteria." *Mar. Biotechnol.* 1:59-65.

Olsen, D.A., D.A. Nellis, and R.S. Wood. 1984. "Ciguatera in the eastern Caribbean." *Mar. Fish. Rev.* 46:13-18.

Paerl, H.W. 1988. "Nuisance phytoplankton blooms in coastal, estuarine, and inland waters." *Limnol. Oceanogr.* 33(4, part 2):823-847.

Paille, D., C. Hackney, L. Reily, M. Cole, and M. Kilgen. 1987. "Seasonal variation in the fecal coliform population of Louisiana oysters and its relationship to microbiological quality." *Journal of Food Protection.* 50:545-549.

Pan American Health Organization (PAHO). 1988. Hurricane Gilbert—Disaster Reports Series.

Pan American Health Organization (PAHO). 1998a. "Implementing decentralization and protecting the poor." Concept paper by HDD/HDP/PAHO.

Pan American Health Organization (PAHO). 1998b. "Health impact of the Southern Oscillation (El Niño)." Official Document CSP25/10 (11 July 1998).

Pan American Health Organization (PAHO). 1998c. "Disasters, preparedness and mitigation in the Americas." *PAHO Newsletter.* Issue no. 73 (July 1998).

Parkes, R.J., B.A. Cragg, S.J. Bale, J.M. Getliff, K. Goodman, P.A. Rochelle, J.C. Fry, A.J. Weightman, and S.M. Harvey. 1994. "Deep bacterial biosphere in Pacific Ocean sediments." *Nature* (Lond.) 371:410-413.

Passaglia, C., F. Dodge, E. Herzog, S. Jackson, and R. Barlow. 1997. "Deciphering a neural code for vision." *Proc. Natl. Acad. Sci. U.S.A.* 94:12649-12654.

Pathirana, C., P.R. Jensen, and W. Fenical. 1992. "Marinone and debromomarinone, antiobiotic sesquiterpenoid naphthoquinones of a new structure class from a marine bacterium." *Tetrahedron Lett.* 33:7663-7666.

Patz, J.A., K. Strzepek, S. Lele et al. 1998a. "Predicting key malaria transmission factors, biting and entomologic inoculation rates, using modeled soil moisture in Kenya." *J. Trop. Med. International Health.* 3:818-827.

Patz, J.A., W.J.M. Martens, D.A. Focks, and T.H. Jetten. 1998b. "Dengue fever epidemic potential as projected by general circulation models of global climate change." *Environ. Health Perspect.* 106:147-153.

Paul, J.H., J.B. Rose, S.C. Jiang, P. London, X. Xhou, and C. Kellogg. 1997. "Coliphage and indigenous phage in Mamala Bay, Hawaii." *Applied and Environmental Microbiology.* 63:133-138.

Paul, V.J. 1992. "Chemical Defenses of Benthis Marine Invertebrates." *Ecological Roles of Marine Natural Products.* Paul, V.J., ed. Ithaca, New York: Comstock Publishing. pp. 164-188.

Pawlik, J.R. 1993. "Marine Invertebrate Chemical Defenses." *Chem. Rev.* 93:1911-1922.

Pearce, D.W., W.R. Cline, A.N. Achanta, S. Fankhauser, R.K. Pachauri, R.S.J. Tol, and P. Vellinga. 1995. "The social costs of climate change: Greenhouse damage and the benefits of control." *Climate Change 1995: Economic and Social Dimensions of Climate Change.* Contribution of Working Group III to the Second Assessment Report of the Intergovernmental Panel on Climate Change. J.P. Bruce, H. Lee, and E.F. Haites, eds. New York, NY: Cambridge University Press. pp. 195-198.

Perlwitz, J. and H-F. Graf. 1995. "The statistical connection between tropospheric and stratospheric circulation of the Northern Hemisphere in winter." J. Climate. 8:2281-2295.

Pettit, G.R., C.L. Herald, D.L. Doubek, and D.L. Herald. 1982. "Isolation and structure of bryostatin 1." *Am. Chem. Soc.* 104:6846-6848.

Philip, P.A., D. Rea, P. Thavasu, J. Carmichael, N.S. Stuart, H. Rockett, D.C. Talbot, T. Ganesan, G.R. Pettit, F. Balkwill, et al. 1993. "Phase I study of bryostatin 1: Assessment of interleukin 6 and tumor necrosis factor alpha induction in vivo. The Cancer Research Campaign Phase I Committee." *J. Natl. Cancer Inst.* 85(22):1812:1818.

Pielke, R.A. and C.W. Landsea. 1998. "Normalized hurricane damages in the United States: 1925-95." *Weather and Forecasting.* 13:621-631.

Pielke, Jr., R. A. and R. A. Pielke, Sr.. 1997. *Hurricanes: Their Nature and Impacts on Society.* New York, NY: John Wiley & Sons. 279 pp.

Pines, J. 1996. "Cyclin from sea urchins to HeLas: making the human cell cycle." *Biochem. Soc. Trans.* 24:15-33.

Poli, M.A., T.J. Mende, and D.G. Baden. 1986. "Brevetoxins, unique activators of voltage-sensitive sodium channels, bind to specific sites in rat brain synaptosomes." *Mol. Pharmacol.* 30(2):129-135.

Powell, M.D. and S.H. Houston. 1996. "Hurricane Andrew's landfall in south Florida, Part II: Surface wind fields and potential real-time applications." *Wea. Forecasting.* 11:329-349.

Quilliam, M.A., M.W. Gilgan, S. Pleasance, A.S.W. Defreitas, D. Douglas, L. Fritz, T. Hu, J.C. Marr, C. Smyth, and J.L.C. Wright. 1993. "Confirmation of a incident of diarrhetic shellfish poisoning in eastern Canada." pp. 547-552, in *Toxic Phytoplankton Blooms in the Sea.* T.J. Smayda and Y. Shimizu, eds. Amsterdam: Elsevier Science Publishers BV. 952 pp.

Raftos, D.A. 1994. "Allorecognition and humoral immunity in tunicates." *Primordial Immunity: Foundations for the vertebrate immune system.* G. Beck, G.S. Habicht, E. L. Cooper, and J. J. Marchanlonis, eds. *Ann. N. Y. Acad. Sci.* 712:227-244.

Rahmstorff, S. 1995. "Bifurcations of the Atlantic thermohaline circulation in response to changes in the hydrologic cycle." *Nature.* 378:145-149.

Rinehart, K.L., T.G. Holt, N.L. Fregeau, J.G. Stroh, P.A. Keifer, F. Sun, H.L. Li, and D.G. Martin. 1990. "Ecteinascidin-729, 743, 745, 759a, 759b, and 770: Potent antitumor agents from the Caribbean tunicate *Ecteinascidia turbinata.*" *J. Org. Chem.* 55:4512-4515.

Roederer, M., S. De Rosa, R. Gerstein, M. Anderson, M. Bigos, R. Stovel, T. Nozaki, D. Parks, L. Herzenberg, and L. Herzenberg. 1997. "8 color, 10-parameter flow cytometry to elucidate complex leukocyte heterogeneity." *Cytometry.* 29(4):328-339.

Rogers, D.J. and M.J. Packer. 1993. "Vector-borne diseases, models, and global change." *The Lancet.* 342:1282-1284.

Rosas, I., E. Salinas, A. Yel, E. Calva, C. Esalva, and A. Cravioto. 1997. "*Escherica coli* in settled-dust and air samples collected in residential environments in Mexico City." *Applied and Environmental Microbiology.* 63:4093-4095.

Roszak, D.B. and R.R. Colwell. 1987. "Survival strategies of bacteria in the natural environment." *Microbiological Reviews.* 51:365-379.

Roussis, V., W.U. Zhongde, W. Fenical, S.A. Strobel, G.D. Van Duyne, and J. Clardy. 1990. "New anti-inflammatory Pseudopterosins from the marine octocoral *Pseudopterogorgia elisabethae.*" *J. Org. Chem.* 55:4916-4922.

Ruepp, B., K.M. Bohren, and K.H. Gabbay. 1996. "Characterization for the osmotic response element of the human aldose reductase gene promoter." *Proc. Natl. Acad. Sci. U.S.A.* 93:8624-8629.

Rutter, G.A., J.M. Theler, M. Murgia, C.B. Wollheim, T. Pozzan, and R. Rizzuto. 1993. "Stimulated Ca2+ influx raises mitochondrial free Ca2+ to supramicromolar levels in a pancreatic beta-cell line. Possible role in glucose and agonist-induced insulin secretion." *J. Biol. Chem.* 268(30):22385-22390.

Sakamoto, Y., R.F. Lockey, and J.J. Krzanowski. 1987. "Shellfish and fish poisoning related to toxic dinoflagellates." *South. Med. J.* 80:866-872.

Sargent, W. 1987. *The Year of the Crab: Marine Models in Modern Medicine.* New York: W. W. Norton and Company.

Schaeffer, B. 1987. "Deuterstome monophyly and phylogeny." *Evolutionary Biology.* M.K. Hecht, B. Wallace, and G.T. Grance, eds. New York: Plenum Press. Vol. 21:179-235.

Schantz, E.I., J.D. Mold, D.W. Stranger, J. Shavel, F.J. Reil, J.P. Bowden, J.M. Lynch, R.S. Wyler, B. Riegel, and H. Sommer. 1957. *J. Am. Chem. Soc.* 79:5230-5235.

Schantz, E.J. 1984. *Seafood Toxins.* E.P. Ragelis, ed. ACS Symposium Series 262:99-111.

Schluter, S.F., R.M. Bernstein, and J.J. Marchalonis, J.J. 1997. "Molecular origins and evolution of immunoglobulin heavy chain genes of jawed vertebrates." *Immunology Today.* 18(11):543-548.

Schmale, M.C., M.R. Aman, and K.A. Gill. 1996. "A retrovirus isolated from cell lines derived from neurofibromas in bicolor damselfish (*Pomacentrus partitus*)." *J. Gen. Virol.* 77 (Pt. 6):1181-7.

Schmitz, W.J. 1996. "On the world ocean circulation: Volume II, The Pacific and Indian Oceans." Woods Hole Oceanographic Institution Technical Report. WHOI-96-08.

Scofield, V. 1997. "Sea squirt immunity: The AIDS connection." *MBL Science.*

Scofield, V.L., R. Clisham, L. Bandyopadhyay, P. Gladstone, L. Zamboni, and R. Raghupathy. 1992. "Binding of sperm to somatic cells via HLA-DR. Modulation by sulfated carbohydrates." *Journal of Immunology.* 148:1718-1724.

Senderowicz, A.M., G. Kaur, E. Sainz, C. Lang, W.D. Inman, J. Rodriguez, P. Crews, L. Malspeis, M.R. Grever, E.A. Sausville, and K.L.K. Duncan. 1995. "Jasplakinolide's inhibition of the growth of prostate carcinoma cells in vitro with disruption of the actin cytoskeleton." *J. Natl. Cancer Inst.* 87(1):46-51.

Shay, L.K., P.G. Black, A.J. Mariano, J.D. Hawkins, and R.L. Elsberry. 1992. "Upper ocean response to hurricane Gilbert." *J. Geophys. Res.* 97(12):20,227-20,248.

Shay, L.K., G.J. Goni, and P.G. Black. 1998. "Effects of a warm core eddy on Hurricane Opal." *Mon. Wea. Rev.* In press.

Sheets, R.C. 1990. "The National Hurricane Center-Past, present, and future." *Wea. Forecasting.* 5:185-232.

Shumway, S.E. 1990. "A review of the effects of algal blooms on shellfish and aquaculture." *J. World Aquaculture Soc.* 21(2): 65-104.

Silva, P., R.J. Solomon, and F.H. Epstein. 1996. "The rectal gland of *Squalus acanthias*, a model for the transport of chloride. *Kidney International.* 49:1552-1556.

Silva, A.J., J.H. Kogan, P.W. Frankland, and S. Kida. 1998. "CREB and memory." *Annu. Rev. Neurosci.* 21:127-148.

Smayda, T.J. 1990. "Novel and nuisance phytoplankton blooms in the sea: Evidence for a global epidemic." *Toxic Marine Phytoplankton.* E. Graneli, B. Sundstron, L. Edler and D.M. Anderson, eds. New York, NY: Elsevier Sci. Publ. pp. 29-40.

Smith, C.G. and S.I. Music. 1998. "*Pfiesteria* in North Carolina: The medical inquiry continues." *N. C. Med. J.*

Smith, L. C. and E. H. Davidson. 1994. "The echinoderm immune system. Characters shared with vertebrate immune systems and characters arising later in deuterostome phylogeny." *Primordial Immunity: Foundations for the vertebrate immune system.* G. Beck, G.S. Habicht, E. L. Cooper, and J. J. Marchanlonis, eds. *Ann. N. Y. Acad. Sci.* 712:213-226.

Sohn, G. and C. Sautter. 1991. "R-phycoerythrin as a fluorescent label for immunolocalization of bound atrazine residues." *J. Histochem. Cytochem.* 39:921-926

Sommer, A. and W.H. Mosley. 1972. "East Bengal cyclone, Nov. 1970: Epidemiological approach to disaster assessment." *Lancet.* May 13, 1972. pp. 1029-1036.

Sommer, A. and W.H. Mosley. 1973. *The Cyclone: Medical Assistance and Determination of Relief and Rehabilitation Needs. Disaster in Bangladesh. Health crises in a developing nation.* Lincoln C. Chen, ed. Oxford, England: Oxford University Press. 122 pp.

Stauber, R.H., K. Horie, P. Carney, E.A., Hudson, N.I. Tarasova, G.A. Gaitanaris, and G.N. Pavlakis. 1998. "Development and applications of enhanced green fluorescent protein mutants." *BioTechniques.* 24:462-471.

Steidinger, K.A. 1983. "A re-evaluation of toxic dinoflagellate biology and ecology." *Prog. Phycol. Res.* F.E. Round and V.J. Chapman, eds. Elsevier, New York. Vol. 2:147-188.

Steidinger, K.A. and D.G. Baden. 1984. *Dinoflagellates.* D.L. Spector, ed. Orlando, Fla.: Academic Press. pp. 201-261.

Steidinger K.A., J.H. Landsberg, and E.W. Truby. In press. "*Cryptosperidenopsis brodyi* gen. et sp. nov. (Dinophyceae), a small lightly armoured dinoflagellate similar to *Pfiesteria*."

Stone, G.W. and C.W. Finkl. 1995. "Preface to special issue on Hurricane Andrew." *J. Coastal Res. Special Issue.* 21:1-364.

Stone, R. 1995. "If the mercury soars, so may health hazards." *Science.* 267:957-958.

Suffness, M., D.J. Newman, and K. Snader. 1989. *Bioorganic Marine Chemistry.* P.J. Scheuer, ed. New York, NY: Springer-Verlag. Vol. 3:131-168.

Tabacco, M.B., M. Uttamlal, M. McAllister, and D.R. Walt. 1998. "An autonomous sensor and telemetry system for low-level cCO_2 measurements in seawater." Accepted: *Analytical Chemistry.* October, 1998.

Takahashi, C., Y. Takai, Y. Kimura, A. Numata, N. Shigematsu, and H. Tanaka. 1995. "Cytotoxic metabolites from a fungal adherent of a marine alga." *Phytochemistry* 38:155-158.

Takenaka, M., A.S. Preston, H.M. Kwon, and J.S. Handler. 1994. "The tonicity-sensitive element that mediates increased transcription of the betaine transporter gene in response to hypertonic stress." *J. Biol. Chem.* 269:29379-29381.

Talukder, J., G.D. Roy, and M. Ahmad. 1992. "Living with cyclones: study on storm surges prediction and disaster preparedness." Dhaka: Community Development Library.

ter Haar, E., R.J. Kowalski, E. Hamel, C.M. Lin, et al. 1996. "Discodermolide, a cytotoxic marine agent that stabilizes microtubules more potently than taxol." *Biochemistry.* 35:243-250.

Tester, P.A., P.K. Fowler, and J.T. Turner. 1989. "Gulf Stream transport of the toxic red tide dinoflagellate *Ptychodiscus brevis* from Florida to North Carolina." *Novel Phytoplankton Blooms. Causes and Impacts of Recurrent Brown Tides and Other Unusual Blooms.* E.M. Cosper, V.M. Bricelj, and E.J. Carpenter, eds. New York, N.Y.: Springer Verlag. pp. 349-358.

Tester, P.A. and P.K. Fowler 1990. "Brevetoxin contamination of *Mercenaria mercenaria* and *Crassostrea virginica*: A management issue." *Toxic Marine Phytoplankton.* E. Graneli, B. Sundstron, L. Edler, and D.M. Anderson, eds. New York, NY: Elsevier Sci. Publ. pp. 499-503.

Tester, P.A., R.P. Stumpf, F.M. Vukovich, P.K. Fowler, and J.T. Turner. 1991. "An expatriate red tide: Transport, distribution, and abundance." *Limnol. Oceanogr.* 36:1051-1061.

Thomas, C. and S. Scott. 1997. "All stings considered: First aid and medical treatment of Hawaii's marine injuries" University of Hawaii Press. [Online]. http://www.aloha.com/~lifeguards/alsting1.html [Sept. 25, 1998].

Thompson, D.W.J., and J.M. Wallace. 1998. "The Arctic Oscillation signature in the wintertime geopotential height and temperature fields." *Geophys. Res. Lett.* 25:1297-1300.

Towle, D.W., M.E. Rushton, D. Heidysch, J.J. Magnani, M.J. Rose, A. Amstutz, M.K.Jordan, D.W. Shearer, and W.-S. Wu. 1997. "Sodium/proton antiporter in the euryhaline crab *Carcinus maenas*: Molecular cloning, expression and tissue distribution." *Journal of Experimental Biology.* 200: 1003-10014..

Trischman, J.A., P.R. Jensen, W. Fenical. 1994a. "Halobacillin, a cytotoxic cyclic acylpeptide of the iturin class produced by a marine bacillus." *Tetrahedron Lett.* 35:5571-5574.

Trischman, J.A., D.M. Tapiolas, P.R. Jensen, R. Dwight, T.C. McKee, C.M. Ireland, T.J. Stout, J. Clardy. 1994b. "Salinamides A and B: Anti-inflammatory depsipeptides from a marine streptomycete." *J. Am. Chem. Soc.* 116:757-758.

Turner, J.T. and P.A. Tester. 1989. "Zooplankton feeding ecology: Copepod grazing during an expatriate red tide." *Novel Phytoplankton Blooms. Causes and Impacts of Recurrent Brown Tides and Other Unusual Blooms.* E.M. Cosper, V.M. Bricelj, and E.J. Carpenter, eds. New York, NY: Springer Verlag. pp. 359-374.

Turner, J.T. and P.A. Tester. 1997. "Toxic marine phytoplankton, zooplankton grazers, and pelagic food webs." *Limnology and Oceanography.* 42:1203-1214.

Turner, J.T., P.A. Tester, and P.J. Hansen. 1998. "Interactions between toxic marine phytoplankton and metazoan and protistan grazers." *Physiological Ecology of Harmful Algal Blooms.* D.M. Anderson, A.D. Cembella, G.M. Hallegraeff, eds. Berlin: Springer. pp. 453-474.

Tyler, M.A. and H.H. Seliger. 1989. "Time scale variations of estuarine stratifacation parameters and impact on the food chains of Chesapeake Bay." *Estuarine Circulation.* B.J. Neilson, A. Kuo, and J. Abrubaker, eds. New Jersey: Humana Press. 377 pp.

United Nations (UN). 1998. "Honduras Situation Report No. 7." UN Hurricane Mitch Information Center. December 3, 1998. [Online]. Available: http://www.un.hn/mitch/unhsitrep7.html [December 17, 1998].

U.S. Agency for International Development (USAID). 1998. "Central America—Hurricane Mitch Fact Sheet # 21." December 4, 1998. [Online]. Available: http://www.reliefweb.int [December 17, 1998].

U.S. Department of Health and Human Services (DHHS). 1998. "Rotavirus vaccine."

U.S. Food and Drug Administration (FDA). 1996. "Sanitation of shellfish growing areas." *National Shellfish Sanitation Program Manual of Operations Part I.* Washington, D.C.

U.S. Food and Drug Administration (FDA). 1997. "Ciguatera." *Bad Bug Book: Foodborne Pathogenic Microorganisms and Natural Toxins Handbook.* [Online]. Available: http://www.cfsan .fda.gov/~mow/chap36.html [December 31, 1997].

U.S. Food and Drug Administration (FDA). 1998. "Consumers cautioned on raw oysters from Galveston Bay." Press release, July 1,1998. [Online]. Available: http://www.fda.gov/bbs/ topics/news/new00645.html [Sept. 25, 1998]

U.S. Geological Survey (USGS). 1998. "Descriptive model of the July 17, 1998 Papua New Guinea tsunami." [Online]. Available: http://walrus.wr.usgs.gov/docs/tsunami/PNG.html#anchor 1298655 [Sept. 16, 1998]

Valle-Levinson, A. and K.M.M.Lwiza. 1995. "The effects of channels and shoals on exchange between the Chesapeake Bay and the adjacent ocean." *J. Geophys. Res.* 100:18551-18563.

Viso, A.C., D. Pesando, and C. Baby. 1987. "Antibacterial and antifungal properties of some marine diatoms in culture." *Bot. Mar.* 30:41-45.

Walsh, P., B. Tucker, and T. Hopkins. 1994. "Effects of confinement/crowding on ureogenesis in the gulf toadfish *Opsanus beta.*" *J. Exp. Biol.* 191:195-206.

Watkins, W.D., and V.J. Cabelli. 1985. "Effect of fecal pollution on *Vibrio parahaemolyticus* densities in an estuarine environment." *Applied and Environmental Microbiology.* 49:1307-1313.

West, P.A., G.C. Okpokwasili, P.R. Brayton, D.J. Grimes, and R.R. Colwell. 1984. "Numerical taxonomy of phenanthrene-degrading bacteria isolated from the Chesapeake Bay." *Applied and Environmental Microbiology.* 48:988-993.

Williams, J. 1962. *Oceanography, An Introduction to the Marine Sciences.* Boston, Massachusetts: Little, Brown and Company.

Winston, J.E. 1988. *Biomedical Importance of Marine Organisms.* D.G. Fautin, ed. San Francisco: California Academy of Sciences. pp. 1-6.

Wood, A.M. and L.M. Shipiro, eds. 1993. *Domoic Acid Final Report of the Workshop.* Corvallis, Oregon: Oregon State Univ. Sea Grant. pp. 1-21.

Wood, C.M., K.M. Gilmour, S.F. Perry, P. Part, and P.J. Walsh. 1998. "Pulsatile urea excretion in gulf toadfish (*Opsanus beta*): evidence for activation of a specific facilitated diffusion transport system." *J. Exp. Biol.* 201(Pt. 6):805-817.

Woods Hole Oceanographic Institution (WHOI). (1998). "Expansion of HAB problems in the U.S." [Online]. Available: http://habserv1.whoi.edu/hab/HABdistribution/habexpand.html [Sept. 15, 1998]

World Health Organization (WHO). 1946. Official Records, 2, 100.

World Health Organization (WHO). 1996. "Malaria." WHO Fact Sheet No. 94.

World Health Organization (WHO). 1998. "El Niño and Its Health Impacts." WHO Fact Sheet No. 192.

Wright, A.E., D.A. Forleo, G.P. Gunawardana, S.P. Gunasekera, F.E. Koehn, and O.J. McConnell. 1990. "Antitumor tetrahydroisoquinoline alkaloids from the colonial ascidian *Ecteinascidia turbinata.*" *J. Org. Chem.* 55:4508-4512.

Wright, A.E. and P.J. McCarthy. 1994. "Drugs from the sea at Harbor Branch." *Sea Technology.* 35(8):10-19.

Xu, H.-S., N. Roberts, F.L. Singleton, R.W. Attwell, D.J. Grimes, and R.R. Colwell. 1982. Survival and viability of nonculturable *Escherichia coli* and *Vibrio cholerae* in the estuarine and marine environment." *Microbiology Ecology.* 8:313-323.

Yancey, P. H., M.E. Clark, S.C. Hand, R.D. Bowlus, and G.N. Somero. 1982. "Living with water stress: Evolution of osmolyte systems." *Science.* 217:1214-1222.

Yasumoto, T., Y. Oshima, and W. Sugawara. 1980. "Identification of *Donophysis fortil* as the causative organism of diarrhetic shellfish poisoning." *Bull. Jap. Soc. Sci. Fish.* 46:1405-1411.

Yayanos, A.A. 1995. "Microbiology to 10,500 meters in the deep sea." *Annu. Rev. Microbiol.* 49:777-805.

Zeballos, J.L. 1993. "Effects of natural disasters on the health infrastructure: Lessons from a medical perspective." *Bulletin of PAHO.* 27(4).

Zeppetello, M.A. 1985. "National and international regulation of ocean dumping: A mandate to terminate marine disposal of contaminated sewage sludge." *Ecology Law Quarterly* 12:619.

Zilinskas, R.A., R.R. Colwell, D.W. Lipton, and R.T. Hill. 1995. *The Global Challenge of Marine Biotechnology.* College Park, Maryland: Maryland Sea Grant College. pp. ix, x, 126.

Appendixes

Appendix A

Committee Biographies

WILLIAM FENICAL received his Ph.D. in organic chemistry from the University of California at Riverside in 1968. Since 1983, he has served as a professor of oceanography for Scripps Institution of Oceanography (SIO) at the University of California in San Diego. In 1996, Dr. Fenical took on the role of director of the Center for Marine Biotechnology and Biomedicine at SIO. In addition, he serves as the coordinator for the University of California Sea Grant College Program. Dr. Fenical's background is in the area of marine chemistry.

DANIEL BADEN received his Ph.D. in biochemistry from the University of Miami in 1977 and he currently serves as a professor of marine biology at the University of Miami. Dr. Baden directs one of five NIEHS Marine and Freshwater Biomedical Sciences Centers. Miami's Center focuses on marine toxicology, with an active interest in toxic dinoflagellates and the hazardous environmental chemicals they produce.

MAURICE BURG earned his M.D. from Harvard Medical School in 1955. He currently serves as chief of the Laboratory of Kidney and Electrolyte Metabolism for the National Heart, Lung, and Blood Institute of the National Institutes of Health. Dr. Burg's research in kidney homeostasis has focused on how osmolytes counteract the denaturing effects of urea in the medulla of the kidney, a compensatory mechanism that was first identified in studies of the high concentrations of urea in the tissues of sharks and rays. Dr. Burg is a member of the National Academy of Sciences.

CLAUDE DE VILLE DE GOYET received his M.D. from the University of Louvain, Belgium in 1965. He currently serves as the chief of emergency preparedness for the Pan American Health Organization. Dr. de Ville de Goyet's formal training is in tropical medicine, public hygiene, malariology, and filariology; however, his avocation is the application of this knowledge to disaster relief (specifically, ocean disasters).

DARRELL JAY GRIMES received his Ph.D. in microbiology from Colorado State University in 1971. He currently serves as the director of the Institute of Marine Sciences at the University of Southern Mississippi. Dr. Grimes' research interests include the microbiology of waste disposal and environmental contaminants, microbiological quality of water resources, and the long-term survival of bacteria.

MICHAEL KATZ, a pediatrician, received his M.D. degree in 1956 from the State University of New York and, in 1963, earned a M.S. degree from Columbia University in tropical medicine and parasitology. His clinical expertise is in pediatric infectious diseases. He currently serves as vice president for research for the March of Dimes Birth Defects Foundation and is also Carpentier Professor, Emeritus of Pediatrics at Columbia University, where he chaired the Department of Pediatrics from 1976 to 1992. Dr. Katz's research interests include mechanisms of latency of neurotrophic viruses, relationship between malnutrition and infection, and diarrheal disease. Dr. Katz is a member of the Institute of Medicine.

NANCY MARCUS received a Ph.D. in biology from Yale University in 1976. She currently serves as director of the Florida State University Marine Laboratory and is a professor in the Department of Oceanography. Dr. Marcus' research interests include evolution, ecology, population genetics of marine zooplankton, and dormancy. She is currently a member of the Ocean Studies Board.

SHIRLEY POMPONI earned her Ph.D. in biological oceanography from the University of Miami, RSMAS, in 1977. For the past four years she has led the Harbor Branch Oceanographic Institution's Division of Biomedical Marine Research in the discovery of novel, marine-derived, biologically-active compounds with therapeutic potential. The major emphasis of her research is on the development of methods for sustainable use of marine resources for drug discovery and development.

PETER RHINES received his Ph.D. in oceanography at Trinity College, Cambridge University, in 1967. Dr. Rhines currently serves as a professor of oceanography and atmospheric sciences at the University of Washington. His research interests include the ocean/atmosphere general circulation, climate change, and the motion of trace chemicals. He has a field program in the Labrador Sea, as well as maintaining a geophysical fluid dynamics laboratory and computer modeling. Dr. Rhines is a member of the National Academy of Sciences.

Patricia Tester earned a Ph.D. in oceanography at Oregon State University in 1983. She serves as a research fishery biologist for the National Marine Fisheries Service. Dr. Tester's interests include the effect of climatology, circulation, and water column conditions on the initiation, growth, and transport of phytoplankton blooms.

John Vena earned his Ph.D. in epidemiology from the State University of New York in 1980. He serves as associate chairman and professor for the Department of Social and Preventive Medicine at the University at Buffalo, School of Medicine and Biomedical Sciences. Dr. Vena has a wide variety of research interests in Environmental Epidemiology, which have included risk perception, and the impact of consumption of contaminated fish on reproductive health.

NRC Staff:

Susan Roberts (project director) earned a Ph.D. in Marine Biology from the Scripps Institution of Oceanography. Dr. Roberts is a program officer for the National Research Council's Ocean Studies Board. Dr. Roberts staffs studies on marine resources and health effects of climate change at the National Research Council. Her research interests include marine microbiology, fish physiology, marine biotechnology, and biomedicine.

Appendix B

Acronyms and Abbreviations

ADP	Adenosine 5'-DiPhosphate
AFRES	U.S. Air Force REServe
AIDS	Acquired ImmunoDeficiency Syndrome
APBL	Atmospheric Boundary Layer
AR	Aldose Reductase gene
ARA	ARachidonic Acid
ASP	Amnesic Shellfish Poisoning
AVHRR	Advanced Very High Resolution Radiometer
CDC	Centers for Disease Control and Prevention
CFP	Ciguatera Fish Poisoning
CICR	Calcium-Induced Calcium Release
CLIVAR	Climate Variability and Predictability Programme
CPHC	Central Pacific Hurricane Center
CRADA	Cooperative Research and Development Agreement
DHA	DocosaHexenoic Acid
DNA	Deoxyribonucleic Acid
DOD	U.S. Department of Defense
DSP	Diarrheic Shellfish Poisoning
ELISA	Enzyme-Linked Immunosorbent Assay
EMS	Emergency Medical Services
ENSO	El Niño/Southern Oscillation
EPA	U.S. Environmental Protection Agency
FDA	U.S. Food and Drug Administration
FEMA	Federal Emergency Management Agency
GFP	Green Fluorescent Protein

119

GOOS	Global Ocean Observing System
HABs	Harmful Algal Blooms
HACCP	Hazard Analysis Critical Control Point
HHS	U.S. Department of Health and Human Services
HIV	Human Immuno-deficiency Virus
hPa	hectoPascals
IgM	Immunoglobulin M
JTWC	Joint Typhoon Warning Center
MHCs	Major Histocompatability Antigens
MPF	Meiosis Promoting Factor
MPH	Miles Per Hour
NAADP	Nicotinic Acid Adenine Dinucleotide Phosphate
NAO	North Atlantic Oscillation
NCI	National Cancer Institute
NHC	National Hurricane Center
NIH	National Institutes of Health
NOAA	National Oceanic and Atmospheric Administration
NPNCDDGs	Natural Products National Cancer Drug Discovery Groups
NSF	National Science Foundation
NSP	Neurotoxic Shellfish Poisoning
NSSP	National Shellfish Sanitation Program
OBL	Ocean Boundary Layer
ONR	Office of Naval Research
ORE	Osmotic Response Element
PAH	Polycyclic Aromatic Hydrocarbons
PAHO	Pan American Health Organization
PCR	Polymerase Chain Reaction
PPM	Parts Per Million
PPT	Parts Per Thousand
PSP	Paralytic Shellfish Poisoning
PUFAs	PolyUnsaturated Fatty Acids
RNA	Ribonucleic Acid
RT-PCR	Reverse Transcriptase Polymerase Chain Reaction
RVF	Rift Valley Fever
SBIR	Small Business Innovative Research
SeaWiFS	Sea-viewing Wide Field-of-view Sensor
SRA	Scanning Radar Altimeter
SST	Sea Surface Temperature
TCs	Tropical Cyclones
TOPEX	The Ocean TOPography Experiment
USAID	U.S. Agency for International Development
USWRP	U.S. Weather Research Program
WAMDI	WAve Model Development and Implementation Group

WCRs   Warm Core Rings
WHO   World Health Organization
WWRP   World Weather Research Program

Appendix C

Workshop Program

Ocean Studies Board
Committee on The Ocean's Role in Human Health
American Geophysical Union (AGU) Building
2000 Florida Avenue, N.W., Washington, DC
June 8-10, 1998

PROGRAM

Monday, June 8 - Conference Room A

9:00 a.m.   Welcome/Introductions - *William Fenical, Chair*

SESSION I: Marine Natural Disasters

9:30 a.m.   "The Role of Ocean Systems in Marine Natural Disasters"
Peter Rhines (University of Washington)

10:00 a.m.   "Estuarine and Near-Shore Transport and Stratification"
William Wiseman (Louisiana State University)

10:30 a.m.   BREAK

10:45 a.m.   "Oceanic Processes Excited by Hurricanes"
Nick Shay (University of Miami, RSMAS)

11:15 a.m.   "Public Health Impacts of Tropical Storms and El Niño"
Claude de Ville de Goyet (Pan American Health Organization)

11:45 a.m.   Discussion Session

12:00 p.m.   LUNCH BREAK

SESSION II: Infectious Diseases

1:00 p.m.   "Waterborne Diseases: Are They Really Detectable, Predictable, and Preventable?"
            D. Jay Grimes (The University of Southern Mississippi)

1:30 p.m.   "Climatic Events Associated with Viral Contamination of Estuaries"
            Joan Rose (University of South Florida)

2:00 p.m.   "Global Warming and Potential Changes in Vector-borne Diseases"
            Milan Trpis (School of Public Health, Johns Hopkins University)

2:30 p.m.   "Infectious Diseases in the Developing World"
            Frances Carr (USAID)

3:00 p.m.   Discussion Session

3:15 p.m.   BREAK

SESSION III: Harmful Algal Blooms

3:30 p.m.   "Harmful Algal Blooms & Marine Food Webs"
            Pat Tester (National Marine Fisheries Service, NOAA)

4:00 p.m.   "Marine Toxins: Orphan Receptors, Charlatan Regulators, and Emerging and Re-Emerging Disease"
            Dan Baden (University of Miami)

4:30 p.m.   "Marine Toxins and Public Health"
            Lorraine Backer (CDC: National Center for Environmental Health)

5:00 p.m.   "Natural Toxins in Seafood: Challenges and Possibilities"
            Sherwood Hall (Food and Drug Administration)

5:30 p.m.   Discussion Session

5:45 p.m.   Reception

6:45 p.m.   Adjourn for the day.

Tuesday, June 9 - Conference Room A

SESSION IV: Marine Organisms as Models for Biomedical Research

9:00 a.m.   "From Sea Urchins to Man: What Marine Embryo Research Tells Us About Human Health and Disease"
            David Epel (Stanford University)

9:30 a.m.   "Evolutionarily Conserved Mechanisms of Salt Tolerance: Marine
             Organisms and the Human Kidney"
             Joan Ferraris (National Institutes of Health)

10:00 a.m.   "Antibodies of Sharks: Evolutionary Emergence and Relevance to
             Autoimmunity and Infection"
             John Marchalonis (University of Arizona)

10:30 a.m.   BREAK

11:00 a.m.   "An Evolutionary and Genetic Approach for Understanding the
             Neurobiological Basis of Vertebrate Behavior"
             Robert Baker (New York University Medical School)

11:30 a.m.   Discussion Session

11:45 a.m.   Lunch Break

SESSION V: Marine Natural Products

1:00 p.m.   "Marine Microorganisms: Developing a New Drug Resource"
             William Fenical (Scripps Institution of Oceanography)

1:30 p.m.   "Marine Natural Products: Therapeutic Potential of Compounds
             Derived from Marine Invertebrates"
             Shirley Pomponi (Harbor Branch Oceanographic Institution,
             Inc.)

2:00 p.m.   "Learning Drug Design from Marine Snails"
             Baldomero Olivera (University of Utah)

2:30 p.m.   Discussion Session

2:45 p.m.   BREAK

3:00 p.m.   Panel Discussion: "Health, Ecological and Economic Dimensions
             of Global Change: Tracking Marine Disturbance and Disease"
             Ben Sherman, Erika Siegfried (Harvard Medical School, Center
             for Health and the Global Environment)

             "Application of Remote Sensing for the Detection of *Vibrio
             cholerae* by Indirect Measurements"
             Anwarul Huq (University of Maryland)

4:30 p.m.   Adjourn Public Session

Index